Interactive Dynamic-System Simulation

Second Edition

Numerical Insights

Series Editor
A. Sydow, GMD-FIRST, Berlin, Germany

Editorial Board
P. Borne, École de Lille, France; G. Carmichael, University of Iowa, USA;
A. Iserles, University of Cambridge, UK; A. Jakeman, Australian National University,
Australia; G. Korn, Industrial Consultants (Wenatchee, WA), USA; G.P. Rao, Indian
Institute of Technology, India; J.R. Rice, Purdue University, USA; A.A. Samarskii, Russian
Academy of Science, Russia; Y. Takahara, Tokyo Institute of Technology, Japan

The Numerical Insights series aims to show how numerical simulations provide valuable insights into the mechanisms and processes involved in a wide range of disciplines. Such simulations provide a way of assessing theories by comparing simulations with observations. These models are also powerful tools which serve to indicate where both theory and experiment can be improved.

In most cases the books will be accompanied by software on disk demonstrating working examples of the simulations described in the text.

The editors will welcome proposals using modelling, simulation and systems analysis techniques in the following disciplines: physical sciences; engineering; environment; ecology; biosciences; economics.

Volume 1
Numerical Insights into Dynamic Systems: Interactive Dynamic System Simulation with Microsoft® Windows™ and NT™
Granino A. Korn

Volume 2
Modelling, Simulation and Control of Non-Linear Dynamical Systems: An Intelligent Approach using Soft Computing and Fractal Theory
Patricia Melin and Oscar Castillo

Volume 3
Principles of Mathematical Modeling: Ideas, Methods, Examples
A.A. Samarskii and A. P. Mikhailov

Volume 4
Practical Fourier Analysis for Multigrid Methods
Roman Wienands and Wolfgang Joppich

Volume 5
Effective Computational Methods for Wave Propagation
Nikolaos A. Kampanis, Vassilios A. Dougalis, and John A. Ekaterinaris

Volume 6
Genetic Algorithms and Genetic Programming: Modern Concepts and Practical Applications
Michael Affenzeller, Stephan Winkler, Stefan Wagner, and Andreas Beham

Volume 7
Interactive Dynamic-System Simulation, Second Edition
Granino A. Korn

Interactive Dynamic-System Simulation

Second Edition

Granino A. Korn

CRC Press
Taylor & Francis Group
Boca Raton London New York

CRC Press is an imprint of the
Taylor & Francis Group, an **informa** business

A CHAPMAN & HALL BOOK

CRC Press
Taylor & Francis Group
6000 Broken Sound Parkway NW, Suite 300
Boca Raton, FL 33487-2742

First issued in paperback 2017

ISBN 13: 978-1-138-11521-7 (pbk)
ISBN 13: 978-1-4398-3641-5 (hbk)

Library of Congress Cataloging-in-Publication Data

Korn, Granino A. (Granino Arthur), 1922-
    Interactive dynamic-system simulation / author, Granino A. Korn. -- 2nd ed.
      p. cm. -- (Numerical insights ; 7)
    "A CRC title."
    Includes bibliographical references and index.
    ISBN 978-1-4398-3641-5 (alk. paper)
    1. Computer simulation. 2. Dynamics. 3. Microcomputers. 4. Interactive computer systems. I. Title. II. Series.

QA76.9.C65K68 2011
004--dc22
                                                               2010014881

Visit the Taylor & Francis Web site at
http://www.taylorandfrancis.com

and the CRC Press Web site at
http://www.crcpress.com

Contents

Preface, xv

CHAPTER 1   Interactive Dynamic-System Simulation   1

1.1   DYNAMIC-SYSTEM MODELS AND SIMULATION
       PROGRAMS   1

    1.1.1   Introduction   1

    1.1.2   Time Histories, State-Transition Models, and
          Differential Equations   3

    1.1.3   Differential Equation Systems with Defined
          Variables   4

1.2   SIMULATION PROGRAMS EXERCISE MODELS   5

    1.2.1   Interpreted Experiment Protocols Call Compiled
          Simulation Runs   5

    1.2.2   Multirun Simulation Studies   5

    1.2.3   What a Simulation Run Does: Updating Variables   6

    1.2.4   What a Simulation Run Does: Output Timing   7

          1.2.4.1   Time-History Sampling   7

          1.2.4.2   Operations with Sampled Data   8

    1.2.5   What a Simulation Run Does: Numerical
          Integration   9

          1.2.5.1   Euler Integration   9

          1.2.5.2   Improved Integration Rules   9

          1.2.5.3   Integration through Discontinuities 11

1.3   HANDS-ON SIMULATION ON THE PC DESKTOP   11

    1.3.1   A Wish List for Interactive Modeling   11

	1.3.2	The Very Simplest Way to Install and Uninstall Programs	12
	1.3.3	Linux versus Windows®	12
	1.3.4	Getting Started: User Programs and Editor Windows	13
		1.3.4.1 Start Desire	13
		1.3.4.2 Enter and Run a Simulation Program	13
	1.3.5	Interactive Modeling with Multiple Editor Windows	14
	1.3.6	A Complete Simulation Program	14
	1.3.7	Time-History Output	17
		1.3.7.1 Programming Time-History Graphs and Listings	17
		1.3.7.2 Display Scaling and Stripchart-Type Displays	17
		1.3.7.3 Time-History Storage and Printing	18
	1.3.8	Preparing Publication Copy	19
	REFERENCES		19
CHAPTER 2	A Gallery of Simple Simulation Programs		21
2.1	INTRODUCTION		21
	2.1.1	Basics	21
	2.1.2	Experiment-Protocol Programs and Commands	21
	2.1.3	**term** and **if** Statements in DYNAMIC Program Segments	23
2.2	EXAMPLES FROM PHYSICS		23
	2.2.1	Classical Applications and Higher-Order Differential Equations	23
	2.2.2	Nonlinear Oscillators and Phase-Plane Plots	24
		2.2.2.1 Van der Pol's Differential Equation	24
		2.2.2.2 Simulation of a Simple Pendulum	26
		2.2.2.3 Lorenz Differential Equations Produce Chaos	28

	2.2.3	A Simple Nuclear Reactor Simulation	28
	2.2.4	An Electric Circuit Simulation with 1401 Differential Equations	29
2.3	AEROSPACE AND RELATED APPLICATIONS		32
	2.3.1	Ballistic Trajectories	32
	2.3.2	Simple Flight Simulation	35
		2.3.2.1 Pitch-Plane Flight Equations	35
		2.3.2.2 Linearized Flight Equations	36
	2.3.3	A Simplified Autopilot	37
	2.3.4	Torpedo Trajectory	38
	2.3.5	Translunar Satellite Orbit	41
2.4	MODELING POPULATION DYNAMICS		43
	2.4.1	Simulation of Epidemic Propagation	43
	2.4.2	Simulation in Ecology: A Host–Parasite Problem	45
		2.4.2.1 A Host–Parasite Problem	45
		2.4.2.2 Generalizations	45
	REFERENCES		47

CHAPTER 3   Introduction to Control System Simulation   49

3.1	SIMULATION AND CONTROL SYSTEM DESIGN		49
	3.1.1	Introduction	49
	3.1.2	Simulation of a Simple Servomechanism	49
	3.1.3	Simulation Studies and Parameter Optimization	51
		3.1.3.1 Test Inputs and Error Measures	51
		3.1.3.2 Parameter-Influence Studies	52
		3.1.3.3 Iterative Parameter Optimization	53
	3.1.4	Where Do We Go from Here?	56
		3.1.4.1 More Elaborate Controllers	56
		3.1.4.2 More Elaborate Plant Models and Control System Noise	56
		3.1.4.3 Control System Transfer Functions and Frequency Response	56
3.2	DEALING WITH SAMPLED DATA		57

3.2.1   Models Using Difference Equations   57

3.2.2   Sampled-Data Operations   58

3.2.3   Changing the Sampling Rate   59

3.3   DIFFERENCE EQUATION PROGRAMMING   59

3.3.1   Primitive Difference Equations   59

3.3.2   General Difference Equation Systems   60

3.3.3   Combined Systems Imply Sample/Hold Operations   61

     3.3.3.1   *Difference Equation Code and Differential Equation Code*   61

     3.3.3.2   *Transferring Sampled Data*   64

     3.3.3.3   *Simulation of Sampled-Data Reconstruction*   64

3.4   A SAMPLED-DATA CONTROL SYSTEM   64

3.4.1   Simulation of an Analog Plant with a Digital PID Controller   64

REFERENCES   67

CHAPTER 4   Function Generators and Submodels   69

4.1   OVERVIEW   69

4.1.1   Introduction   69

4.2   GENERAL-PURPOSE FUNCTION GENERATION   69

4.2.1   Library Functions   69

4.2.2   Function Generators Using Function Tables   70

     4.2.2.1   *Functions of One Variable*   70

     4.2.2.2   *Functions of Two Variables*   71

     4.2.2.3   *General Remarks*   72

4.2.3   User-Defined Functions   73

4.3   LIMITERS AND NONCONTINUOUS FUNCTIONS   74

4.3.1   Limiters   74

     4.3.1.1   *Introduction*   74

     4.3.1.2   *Simple Limiters*   74

4.3.1.3	*Useful Relations between Limiter Functions*	75
4.3.1.4	*Maximum and Minimum Functions*	76
4.3.1.5	*Output-Limited Integrators*	76
4.3.2	Switches and Comparators	76
4.3.3	Signal Quantization	77
4.3.4	Noise Generators	77
4.3.5	Integration through Discontinuities and the Step Operator	79
4.4	**VERY USEFUL MODELS EMPLOY SIMPLE RECURRENCE RELATIONS**	**80**
4.4.1	Introduction	80
4.4.2	Track/Hold Circuits and Maximum/Minimum Tracking	81
4.4.3	Models with Hysteresis	83
4.4.3.1	*Simple Backlash and Hysteresis*	83
4.4.3.2	*A Comparator with Hysteresis*	83
4.4.3.3	*A Deadspace Comparator with Hysteresis*	84
4.4.4	Signal Generators	84
4.4.4.1	*Square Wave, Triangle, and Sawtooth Waveforms*	84
4.4.4.2	*Signal Modulation*	87
4.4.5	Generation of Inverse Functions	88
4.5	**SUBMODELS CLARIFY SYSTEM DESIGN**	**89**
4.5.1	Submodel Declaration and Invocation	89
4.5.1.1	*Submodels*	89
4.5.1.2	*Submodel Declaration*	89
4.5.1.3	*Submodel Invocation and Invoked State Variables*	90
4.5.2	A Simple Example: Coupled Oscillators	90
4.6	**A BANG-BANG CONTROL SYSTEM SIMULATION USING SUBMODELS**	**92**
4.6.1	A Satellite Roll-Control Simulation	92

4.6.2 Bang-Bang Control and Integration 95

REFERENCES 97

CHAPTER 5 ▪ Programming the Experiment Protocol 99

5.1 INTRODUCTION 99

5.2 PROGRAM CONTROL 99

5.2.1 Labels and Branching 99

5.2.2 Conditional Branching 100

5.2.3 **for, while,** and **repeat** Loops 101

5.2.4 Experiment-Protocol Procedures 101

5.3 ARRAYS AND SUBSCRIPTED VARIABLES 103

5.3.1 Arrays, Vectors, and Matrices 103

5.3.1.1 *Simple Array Declarations* 103

5.3.1.2 *Equivalent Arrays* 104

5.3.1.3 **STATE** *Arrays* 105

5.3.2 Filling Arrays with Data 105

5.3.2.1 *Simple Assignments* 105

5.3.2.2 **data** *Lists and* **read** *Assignments* 106

5.3.2.3 *Text-File Input* 107

5.4 EXPERIMENT-PROTOCOL OUTPUT AND INPUT 107

5.4.1 Console, Text-File, and Device Output 107

5.4.1.1 *Console Output* 107

5.4.1.2 *File and Device Output* 108

5.4.1.3 *Closing Files or Devices* 109

5.4.2 Console, File, and Device Input 109

5.4.2.1 *Interactive Console Input* 109

5.4.2.2 *File or Device Input* 109

5.4.3 Multiple DYNAMIC Program Segments 110

5.5 EXPERIMENT-PROTOCOL DEBUGGING, NOTEBOOK FILE, AND HELP FILES 111

5.5.1 Interactive Error Correction 111

5.5.2 Debugging Experiment-Protocol Scripts 112

5.5.3 The Notebook File   112

5.5.4 Help Facilities   113

REFERENCE   113

CHAPTER 6   Models Using Vectors and Matrices   115

6.1 OVERVIEW   115

6.1.1 Introduction   115

6.2 VECTORS AND MATRICES IN EXPERIMENT-PROTOCOL SCRIPTS   116

6.2.1 Null Matrices and Identity Matrices   116

6.2.2 Matrix Transposition   117

6.2.3 Matrix/Vector Sums and Products   117

6.2.4 **MATRIX** Products   117

6.2.5 Matrix Inversion and Solution of Linear Equations   118

6.3 VECTORS AND MATRICES IN DYNAMIC-SYSTEM MODELS   118

6.3.1 Vector Expressions in DYNAMIC Program Segments   118

6.3.2 Matrix/Vector Products in Vector Expressions   119

    6.3.2.1 *Matrix/Vector Products*   119

    6.3.2.2 *Example: Rotation Matrices*   120

6.3.3 Matrix/Vector Models of Linear Systems   121

6.4 VECTOR INDEX-SHIFT OPERATIONS   123

6.4.1 Index-Shifted Vectors   123

6.4.2 Simulation of an Inductance/Capacitance Delay Line   125

6.4.3 Programming Linear-System Transfer Functions   125

    6.4.3.1 *Analog Systems*   125

    6.4.3.2 *Digital Filters*   126

6.5 **DOT** PRODUCTS, SUMS, AND VECTOR NORMS   129

6.5.1 **DOT** Products and Sums of **DOT** Products   129

6.5.2 Euclidean Norms   130

6.5.3 Simple Sums: Taxicab and Hamming Norms   131

6.6   MORE VECTOR/MATRIX OPERATIONS   131

   6.6.1   Vector Difference Equations   131

   6.6.2   Dynamic-Segment Matrix Operations   132

         6.6.2.1   Vector Products   132

         6.6.2.2   Simple **MATRIX** Assignments   132

         6.6.2.3   A More General Technique   133

   6.6.3   Submodels with Vectors and Matrices   133

6.7   MODEL REPLICATION: A GLIMPSE OF ADVANCED
        APPLICATIONS   134

   6.7.1   Model Replication   134

         6.7.1.1   Vector Assignments Replicate Models   134

         6.7.1.2   Parameter-Influence Studies   135

         6.7.1.3   Vectorized Monte Carlo Simulation   137

         6.7.1.4   Neural Network Simulation   137

   6.7.2   Other Applications   138

6.8   TIME HISTORY FUNCTION STORAGE IN ARRAYS   138

   6.8.1   Function Storage and Recovery with **store** and
           get   138

         6.8.1.1   **store** and **get** Operations   138

         6.8.1.2   Application to Automatic Display Scaling   141

   6.8.2   Time Delay Simulation   141

   REFERENCES   144

CHAPTER 7   Modeling Tricks and Treats   145

7.1   OVERVIEW, AND A FIRST EXAMPLE   145

   7.1.1   Introduction   145

   7.1.2   A Benchmark Problem with Logarithmic Plotting   145

7.2   MULTIPLE RUNS CAN SPLICE COMPLICATED TIME
        HISTORIES   146

   7.2.1   Simulation of Hard Impact: The Bouncing Ball   146

   7.2.2   The EUROSIM Peg-and-Pendulum Benchmark [1]   149

   7.2.3   The EUROSIM Electronic-Switch Benchmark   151

7.3   TWO PHYSIOLOGICAL MODELS   153

7.3.1   Simulation of a Glucose Tolerance Test   153

7.3.2   Simulation of Human Blood Circulation   154

7.4   A PROGRAM WITH MULTIPLE DYNAMIC SEGMENTS   159

7.4.1   Crossplotting Results from Multiple Runs: The
Pilot Ejection Problem   159

7.5   FORRESTER-TYPE SYSTEM DYNAMICS   160

7.5.1   A Look at System Dynamics   160

7.5.2   World Simulation   164

REFERENCES   169

CHAPTER 8   General-Purpose Mathematics   171

8.1   INTRODUCTION   171

8.1.1   Overview   171

8.2   COMPILED PROGRAMS NEED NOT BE SIMULATION
PROGRAMS   172

8.2.1   Dummy Integration Simply Repeats DYNAMIC-
Segment Code   172

8.2.2   Fast Graph Plotting   172

8.2.2.1   A Simple Function Plot   172

8.2.2.2   Array-Value Plotting   173

8.2.3   Fast Array Manipulation   174

8.2.4   Statistical Computations   175

8.3   INTEGERS, COMPLEX NUMBERS, AND INTERPRETER
GRAPHICS   176

8.3.1   **INTEGER** and **COMPLEX** Quantities   176

8.3.2   Octal and Hexadecimal Integer Conversions   177

8.3.3   Complex Number Operations and Library
Functions   177

8.3.4   Interpreter Graphics, Complex Number Plots, and
Conformal Mapping   178

8.3.4.1   Interpreter Plots   178

8.3.4.2   Complex Number Plots   178

8.4 FAST FOURIER TRANSFORMS AND
CONVOLUTIONS   180

    8.4.1   Fast Fourier Transforms   180

       8.4.1.1   *Simple Fourier Transformations*   180

    8.4.2   Simultaneous Transformation of Two Real Arrays   180

    8.4.3   Cyclical Convolutions   181

REFERENCE   183

APPENDIX: SIMULATION ACCURACY AND INTEGRATION
TECHNIQUES, 185

A.1   SIMULATION ACCURACY AND TEST PROGRAMS   185

    A.1.1   Introduction   185

    A.1.2   Roundoff Errors   185

    A.1.3   Choice of Integration Rule and Integration Step
Size   186

A.2   INTEGRATION RULES AND STIFF DIFFERENTIAL
EQUATION SYSTEMS   189

    A.2.1   Runge–Kutta and Euler Rules   189

    A.2.2   Implicit Adams and Gear Rules   191

    A.2.3   Stiff Differential Equation Systems   191

    A.2.4   Examples Using Gear-Type Integration   192

    A.2.5   Perturbation Methods Can Improve Accuracy   194

    A.2.6   Avoiding Division   195

REFERENCES   195

INDEX, 197

Preface

COMPUTER SIMULATION IS EXPERIMENTATION with models. I hope that this book will help engineers and scientists use personal computers for modeling and simulation. The book is not a survey or history of simulation programming. It is a hands-on, practical tutorial on interactive dynamic-system modeling and simulation.

The book CD has an industrial-strength simulation-program package.* No installation program is needed. Simply copying a single Windows® or Linux folder from the CD installs a complete, ready-to-run simulation system on a personal computer. Readers can then run and modify every program example in the text, or try their own projects. Our simulation language reads much like ordinary mathematics notation, e.g.,

$$d/dt \; x = -x * \cos(w * t) + 2.22 * a * x$$
$$Vector \; y = A * x + B * u$$

For truly interactive modeling, screen-edited programs are run-time compiled and immediately produce solution displays on a typed run command.

Chapter 1 introduces dynamic-system models and then explains how differential equation–solving software works. Chapter 2 demonstrates real simulation programs with simple examples from physics, aerospace engineering, and population dynamics.

---

* The open-source Desire package admits up to 40,000 differential equations in scalar and/ or vector form, with 15 integration rules and handles fast Fourier transforms and complex frequency-response and root-locus plots. The program readily combines differential equations with difference equations to model sampled-data control systems and neural networks. Desire does not solve differential algebraic equations.

Chapter 3 introduces control-system simulation with a simple servo-mechanism model. We then present a new detailed treatment of the difference equation programs needed to model sampled-data control systems with digital controllers.

The following chapters add programming know-how. Chapter 4 introduces library, table-lookup, and user-definable functions. We go on to limiter, switching, and noise functions and show a simple method for integrating such nondifferentiable functions. The remainder of the chapter discusses the use of submodels for efficient programming.

Every simulation implies an experiment protocol that sets parameters, calls simulation runs, and decides what to do next. Typed commands will do for very simple experiments, but serious studies need a full-fledged programming language to combine and compare results, or to do parameter optimization and statistics. Chapter 5 describes our experiment-protocol scripting language.

Chapter 6 introduces the powerful vector and matrix operations made possible by Desire's novel vectorizing compiler. Compact vector operations can implement model replication: a single simulation run can exercise a thousand models with different parameters or random inputs. Advanced applications of this technique to large parameter-influence studies, vectorized Monte Carlo simulation, and artificial neural networks are described in a separate textbook.*

Chapter 7 describes improved versions of eight classical simulation programs to illustrate a number of useful programming tricks. Finally, Chapter 8 shows how interpreted scripts and compiled DYNAMIC program segments can quickly solve a number of general-purpose mathematical problems, including fast graph plotting, Fourier transformations, and complex-number plots.

The Appendix has hints for selecting integration rules and step size and also introduces perturbation methods.

I am deeply indebted to Theresa M. Korn for her perennial assistance with this and many other projects. I would also like to thank Professor A. Sydow (Fraunhofer/FIRST Institute), Professor H. Vakilzadian (University of Nebraska), Dr. R. Wieland (ZALF/Muencheberg Landscape Research Institute) for years of advice and support, and Professor M. Main (University of Colorado) for valuable help with his Windows graphics library.

Granino A. Korn
Chelan, Washington

* G.A. Korn, *Advanced Dynamic-System Simulation: Model-Replication Techniques and Monte Carlo Simulation*, Wiley, Hoboken, NJ, 2007.

CHAPTER 1

Interactive Dynamic-System Simulation

1.1 DYNAMIC-SYSTEM MODELS AND SIMULATION PROGRAMS

1.1.1 Introduction

We perceive and manipulate real-world objects and relations in terms of simplified *models* that abstract what is currently important. Useful models can involve toy building blocks, sandboxes, or sketches on paper. Engineers and scientists use general-purpose models defined in terms of mathematics and computer programs.

Simulation is experimentation with models. This book deals with dynamic-system models that represent not only systems such as aircraft and power plants but also living organs, populations, and ecosystems. In almost every case, computer-model experiments—simulations—are dramatically cheaper and often safer than real-world experiments. What is more, simulation lets you experiment with the design and operation of projects—say, space vehicles or forest reserves—before they are built. Simulations, of course, manipulate only models and must eventually be validated by real-world tests.

Research and design projects require successive decisions based on many hundreds of simulation runs. This means that we need fast and user-friendly computer programs that permit *interactive modeling*. Interactive experimentation is the specific purpose of the open-source Desire

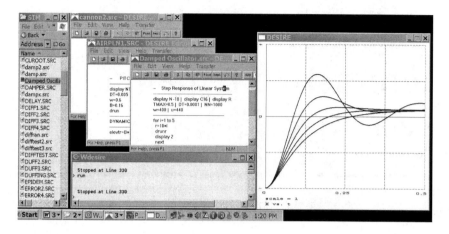

FIGURE 1.1a   Desire running under Microsoft Windows. The dual-screen display shows Command and Graph Windows, a user-program file folder, and three Editor Windows. Programs in different Editor Windows can be run in turn to compare models or different versions of a model. Runtime displays normally show colored curves on a black background.

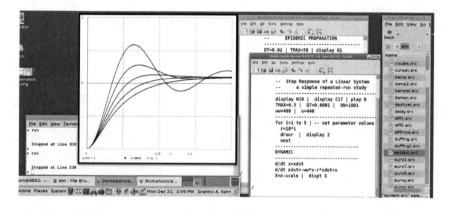

FIGURE 1.1b   The same simulation running under Linux. Here the Command Window is an ordinary Linux terminal window, and the Editor Windows use a standard Linux editor.

simulation package* supplied on the book CD. Figures 1.1a and 1.1b show our user interface. Programs entered and edited in the *Editor Windows* produce output displays in a *Graph Window*. A *Command Window* controls the simulation.

Desire solves differential equations as quickly as a Fortran program. Perhaps more significantly, even large programs are *runtime-compiled* within 0.1 s, so that *solution displays start at once on a typed command*.

This chapter introduces state-transition models of dynamic systems and describes how a differential-equation-solving simulation program does its job. We then exhibit a complete, easily readable simulation program in Section 1.3.6.

1.1.2 Time Histories, State-Transition Models, and Differential Equations

The simplest mathematical models of physical systems are *static models* that relate system variables, say gas pressure p and gas volume V, by equations such as $p = alpha\ V$. *Dynamic-system models* introduce a time variable t and generate time histories of *state variables* $x(t)$ by relating their upcoming values to known past values. A simple form of such a model is a *difference equation* such as

$$\mathbf{x(t + \Delta t) = x(t) + G[t;\ x(t)]} \tag{1.1}$$

$\mathbf{x(t0)}$, successive substitutions produce successive values of $\mathbf{x(t)}$.

Many dynamic-system models in physics, chemistry, and physiology assume that $\mathbf{x(t)}$ is continuous and differentiable. Then, in the limiting case of very small time increments $\mathbf{\Delta t}$, the state equation (1) becomes a first-order *differential equation*

$$\mathbf{dx/dt = G[t;\ x(t)]}$$

---

* Desire stands for Direct Executing SImulation in REal time. The book CD has the complete Windows and Linux versions, source code, a reference manual, and more than 150 application examples. Desire uses double-precision floating-point arithmetic and handles up to 40,000 differential equations in scalar and/or vector form (Chapter 6), plus difference equations (Chapter 3), neural networks, and fuzzy logic [1].

satisfied by the function **x(t)**, again with given initial values **t0** and **x(t0)**.*

Modeling more general dynamic systems requires multiple state variables **x1(t)**, **x2(t)**, ... that satisfy a system of 1st-order differential equations (*state equations*)

$$\textbf{dxi/dt = Gi[t; x1(t), x2(t), ..., xn(t); a1, a1, ...] (i = 1, 2, ..., n)}$$
$$\textbf{(1.2a)}$$

with given initial values **t0** and **xi(t0)**. **a1**, **a1**, ... are constant *model parameters*.

1.1.3 Differential Equation Systems with Defined Variables

More realistic state-transition models introduce functions of the state variables as *defined variables* **yk(t) = Yk(t; x1, x2, ...)**. If, say, the velocity **xdot** of a car is used as a state variable, then the aerodynamic drag

$$\textbf{drag = alpha v}^2$$

is a defined variable used in the state equation

$$\textbf{(d/dt) xdot = propulsion force - drag}$$

alpha is a given system parameter. However, defined variables may also simply be intermediate results introduced for convenience in computations.

More generally, defined variables may depend on other defined variables, that is,

$$\textbf{y1(t) = Yk(t; x2, x2, ... ; y1, y2, ... ; b1, b2, ...)}$$
$$\textbf{y2(t) = Yk(t; x1, x3, ... ; y1, y3, ... ; b1, b2, ...)} \quad \textbf{(1.2b)}$$

where **b1, b2,** ... are more model parameters.

In computer programs, the defined-variable assignments (Equation 1.2b) must precede the derivative assignments (Equation 1.2a). In addition,

---

* Newton's invention of such differential equation models for particle motion simplified time-history prediction so dramatically that it changed physics forever. One needs to specify just one vector function (Newton's "force") to predict any number of different position time histories corresponding to different initial conditions.

the defined-variable assignments (Equation 1.2b) must be sorted into procedural order to permit successive evaluation of each defined variable **yk** in terms of state-variable values computed earlier. Desire returns an error message if that is not possible.*

1.2 SIMULATION PROGRAMS EXERCISE MODELS

1.2.1 Interpreted Experiment Protocols Call Compiled Simulation Runs

Every simulation experiment must have an *experiment protocol* that sets and modifies parameters and calls *simulation runs* to exercise a model. Every Desire program therefore starts with an *experiment-protocol script* that sets model parameters and initial conditions. The program then *defines the model* in a *DYNAMIC program segment* listing the assignments (2) in proper order. DYNAMIC segments also specify time-history output to displays, printers, or files. Both experiment protocol and DYNAMIC program segments use a readable notation much like ordinary mathematics (Section 1.3.6).

Simulation-run code typically executes DYNAMIC-segment assignments hundreds or thousands of times. It must be fast, and is therefore *compiled* at runtime. The less time-critical experiment-protocol script is *interpreted*. Both script and simulation runs execute at once on a typed run command.

1.2.2 Multirun Simulation Studies

Simulation studies manipulate successive simulation run results in possibly complicated programs, say, for parameter optimization or statistical evaluation of results. It follows that effective experiment-protocol scripts need a full-fledged interpreter language with expression assignments, **if** statements, program loops, subroutines, and input/output operations. Desire's script language resembles an advanced Basic dialect augmented with vector, matrix, and complex-number operations and fast Fourier transforms (Chapters 5 and 8). Our experiment protocols can, moreover, use operating-system commands and Unix-type shell

---

* Many simulation programs sort defined-variable assignments automatically. But Desire has a different program feature. Unsortable assignments such as **x1 = x1 + alpha * y** normally return an error message. But if the experiment protocol has assigned an initial value to **x1** then Desire treats **x1 = x1 + alpha * y** as a difference equation that updates **x1** (Chapter 3, Section 3.3.1).

scripts, call external programs, and communicate with them through files or pipes.

1.2.3 What a Simulation Run Does: Updating Variables

As we noted, a differential-equation-solving program lists the model assignments (1,2) in a DYNAMIC program segment. Each call to a *derivative routine* executes these assignments in order. More specifically, the derivative routine computes all derivatives **G1**, **G2**, ..., **Gn** and stores them in a hidden array.

To run a simulation, the experiment protocol initializes displays and other output, and sets parameters and initial values. A programmed or command-mode **drun** statement compiles the derivative routine, if it was not already compiled by an earlier run. **drun** then saves the initial values of simulation time and differential-equation state variables for possible reuse (Section 1.3.6) and starts a simulation run.

Figure 1.2 shows how such a simulation run proceeds. The flow diagram is complicated by the fact that a simulation-run program has to do *two separate jobs.* It must

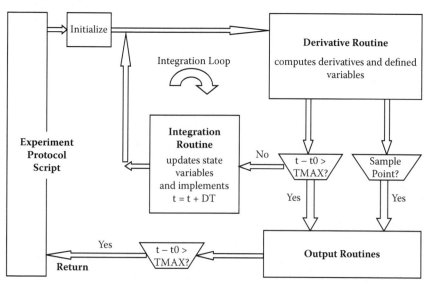

FIGURE 1.2   Desire's differential-equation solver repeats successive fixed or variable integration steps and periodically samples selected variables. An extra set of samples is collected at the end of the simulation run to provide initial values for continued runs.

- Start and stop the run, and produce output at specified communication times
- Update the simulation time and all system variables

An initial call of the derivative routine uses the given initial state-variable values to compute the corresponding initial defined-variable values for $t = t0$. A precompiled *integration routine* selected by the experiment protocol then advances the time variable **t** and again calls the derivative routine, typically several times during each successive integration step (Section 1.2.5). The derivative routine executes the assignments (1,2) to update first all defined variables **yi** and then all derivatives **Gk**. Each assigned expression uses **t** and **xi** values updated by the preceding derivative call, starting with the given initial values.

1.2.4 What a Simulation Run Does: Output Timing

1.2.4.1 Time-History Sampling

Output requests for displays, time-history listings, or file storage such as **dispt X** (Section 1.3.6) are programmed in the DYNAMIC program segment, but they are not part of the derivative routine. Output operations normally execute at **NN** uniformly spaced *communication points* (sampling points)

$$t = t0, t0 + COMINT, t0 + 2\ COMINT, \dots ,$$
$$t0 + (NN - 1)COMINT = t0 + TMAX \qquad (1.3a)$$

where

$$COMINT = TMAX/(NN - 1) \qquad (1.3b)$$

is the *communication interval*. **NN** and **TMAX** are *simulation parameters* set by the experiment-protocol script.

Starting at $t = t0$, the integration routine loops, increasing **t** until **t** reaches the next communication point (3); the solution is then sampled for output. The simulation run terminates after accessing the last sample at $t = t0 + TMAX$, unless the run is stopped earlier, either by the user or by a programmed termination statement (**term**, Chapter 2, Section 2.1.3).

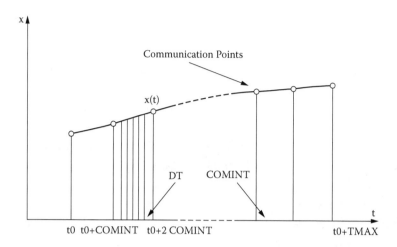

FIGURE 1.3 Time history of a simulation variable **x(t)** with sampling times **t = t0, t0 + COMINT, t0 + 2 COMINT, ... , t0 + TMAX** and integration steps. The integration step at the end of each sampling interval ends on a sampling point (see text).

With reference to Figure 1.3, it would be senseless to sample time-history values computed by an integration routine in the middle of an ongoing integration step. Some simulation programs obtain such time-history values by interpolation between samples outside the integration-step interval, but that slows the simulation. Desire, therefore, simply returns an error message if the sampling interval **COMINT = TMAX/(NN − 1)** is smaller than the integration step **DT**.

Even then, some integration steps could overlap one output-sampling point. To solve this problem, Desire's variable-step integration rules 4 and 8 through 16 (Table A.1 in the Appendix) automatically shorten integration steps when they run into a sampling point. Fixed-step integration rules cannot change **DT**, so they postpone sampling until the integration step completes. That is not noticeable in displays, but it can cause output listings wanted at, say, **t** = 0, 1, 0.2, 0.3, ... to occur instead at slightly modified times like **t** = 0.10003, 0.20005, 0.30002, ... To prevent this, you can select **DT** and **NN** and then set **TMAX** to **(NN-1) * DT**.

1.2.4.2 Operations with Sampled Data

Defined-variable assignments subsequent to an **OUT** *statement execute only at the sampling points* (3) *rather than with every derivative call.* This important feature of the Desire language

- *Saves unnecessary computation of variables needed only for output* (e.g., scaling and shifting displayed curves, logarithmic scales).

- Models *sampled-data operations* in simulated hybrid analog-digital systems, for example, analog plants with digital controllers (Chapter 3).

- Ensures that *pseudorandom-noise generators* are sampled periodically, and never inside an integration step (Chapter 4, Section 4.3.4; and Reference 1).

1.2.5 What a Simulation Run Does: Numerical Integration

1.2.5.1 Euler Integration

Given a differential equation system (1,2), the simplest way to approximate continuous updating of a state variable **xi** in successive integration steps is the *explicit Euler rule* (see also Appendix)

$$\text{xi(t + DT) = xi(t) + Gi[t; x1(t), x2(t), ...; y1(t), y2(t),...; a1, a2, ...] DT}$$
$$\text{(i = 1, 2, ... , n)} \qquad\qquad \text{(1.4)}$$

Gi is the value of **dxi/dt** calculated by the derivative call executing Equation 1.2 at the time **t** (Figure 1.4a). Each integration step ends with a derivative call that produces derivative and defined-variable values for the next step.

1.2.5.2 Improved Integration Rules

Numerical integration has developed for well over 200 years, but is still an active research topic. Voluminous literature and many useful program libraries exist. References 2 and 3 are recent comprehensive treatments of this subject, and the Appendix has more references.

The Euler integration rule (4) simply increments each state variable by an amount proportional to its current derivative (Figure 1.4a). This procedure approximates true integration only for very small integration steps **DT**. *Improved updating requires multiple derivative calls in the course of each integration step* **DT**. This extra computation per step often reduces the number of steps, and thus the total number of derivative calls needed for a specified accuracy. Derivative calls typically account for most of the computing time required by simulation programs.

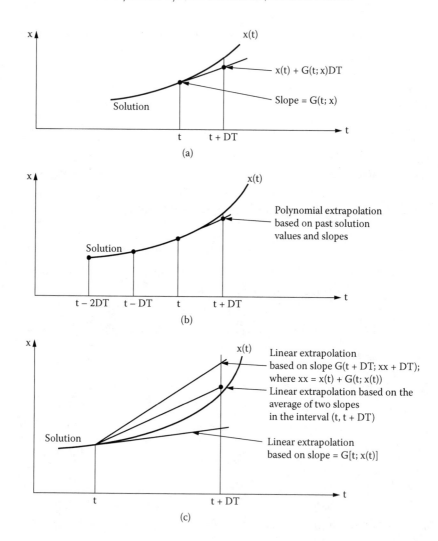

FIGURE 1.4 Numerical integration: (a) Euler rule, (b) a multistep rule, and (c) a Runge–Kutta rule.

- *Multistep rules* extrapolate updated values of the **xi** as polynomials based on **xi** and **Gi** values at several past times **t - DT, t - 2 DT, ...** (Figure 1.4b).

- *Runge–Kutta rules* precompute two or more approximate derivative values in the interval **(t, t + DT)** by Euler-type steps and use their weighted average for updating (Figure 1.4c).

The coefficients in such integration formulas are chosen so that polynomials of degree **N** integrate exactly (**N**th-order integration formula).

Explicit integration rules such as Equation 1.4 express future values **xi(t + DT)** in terms of already computed past state-variable values. *Implicit* rules, such as the *implicit Euler rule*

$$\text{xi(t+DT) = xi(t) + fi[t+DT; x1(t+DT), x2(t+DT), ... ; y1(t+DT), y2(t+DT), ...] DT (i = 1, 2, ..., n)} \qquad (1.5)$$

require a program that *solves the predictor equations (Equation 1.5) for the* **xi(t + DT)** at each step. Implicit rules clearly involve more computation, but they may admit larger **DT** values with acceptable errors (see also Appendix A).

Variable-step integration rules adjust the current integration step sizes **DT** to maintain accuracy estimates that compare various tentative updated solution values. This can save many integration steps and thus much computing time. Variable-step rules are further discussed in the Appendix. Figure A.3 shows how step sizes change in a real application.

The experiment-protocol statement **irule n** selects one of the 15 integration rules listed in Table A.1. The integration rule defaults to Rule 1 (fixed-step 2nd-order Runge–Kutta integration). Section A.4 (in the Appendix) has an interesting program that lets you study and compare the effects of the different integration rules on solution accuracy and speed.

1.2.5.3 Integration through Discontinuities
Integration rules normally assume that all integrands **Gi** are continuous and differentiable over each integration step. This requirement causes serious problems in simulations that involve switches, limiters, and/or noise generators. This topic will be discussed in Chapter 4.

1.3 HANDS-ON SIMULATION ON THE PC DESKTOP

1.3.1 A Wish List for Interactive Modeling
Interactive modeling requires a user-friendly interface that lets you

- Program new models and experiments
- Modify (edit) parameters and models

- View, print, store, and recall program and data files

- Combine existing programs or subprograms

The interface must also make it easy to *run, pause, continue, stop*, and *rerun* simulations and to *observe and store results*. Finally, simulation time histories and error messages must display *during* rather than *after* a simulation run or simulation study. This can save valuable time by letting you stop simulation runs with uninteresting or faulty results.

1.3.2 The Very Simplest Way to Install and Uninstall Programs

A Desire distribution CD contains source-file folders for Windows (**\mydesire**) and for Linux (**\desire**). These folders contain executable programs and subfolders with user-program examples and help files.*

Simply copying **\mydesire** anywhere on a hard disk running Windows immediately produces *a complete ready-to-run modeling/simulation system*. No installation program is needed. Desire can also be run directly from the distribution CD or from a flash stick. One can also run multiple copies of Desire (and thus multiple simulations) concurrently.

Under Linux, we simply copy **\desire** to the hard disk, open a console window, and **cd** to the **\desire** folder.

Uninstallation is equally simple. Uninstallation only requires deleting the installation folder **\mydesire** or **\desire**. No garbage is left on your hard disk.†

1.3.3 Linux versus Windows®

To avoid confusing readers used to the Windows operating system, the remainder of this chapter describes *Windows operations. Linux operations* differ only slightly; the Reference Manual on the book CD has detailed instructions for both operating systems.

Desire is not bothered by the difference between Windows and Linux text-file formats, so that user programs prepared under either system work in both. Only the display colors will differ; you may want to change the color assignments **display N n** and/or **display C m** (Section 1.3.7.1).

---

* These folders are also distributed over the Internet as compressed folders **mydesire.zip** and **desire.tgz** (download from sites.google.com/site/gatmkorn or e-mail).

† Desire avoids the trouble-prone Windows uninstaller and never uses the Windows registry.

That said, we prefer Linux for professional work. The 64-bit Linux version of Desire is substantially faster than the Windows version, because good 64-bit Linux software was available years before comparable Windows systems. Overall, Linux is very reliable and avoids the registry problems that plague Windows after multiple program installations and removals.

1.3.4 Getting Started: User Programs and Editor Windows

1.3.4.1 Start Desire

To *start Desire* under Windows, double-click **Wdesire.bat**, or preferably a shortcut icon on the desktop. This opens the Desire *Command Window,* an *Editor Window,* and the **\mydesire** installation folder. Open one of the user-program subfolders in **\mydesire**, say **\mydesire\sim**, for a selection of working user-program examples.

Size and position these windows as you prefer (Figure 1.1). Use the Command Window's **Properties** menu to fix the Command Window position, size, font, and colors. Desire will then remember the sizes and positions of your Command and Editor windows for future sessions.

1.3.4.2 Enter and Run a Simulation Program

The Desire editor has conventional Windows menus and help screens. Type a new simulation program, or *drag-and-drop an existing user-program file* into the Editor Window. User programs are ordinary text files stored anywhere on your computer, usually with the file extension **.src**.*

Click the red **OK** *button* in the Editor Window to accept the edited program. This program then compiles and *runs immediately* when you activate the Command Window with a mouse click and enter **erun** (or more conveniently **zz**). Programmed displays then appear immediately in the Graph Window.†

A Command Window prompt **>** indicates that a simulation run or simulation study is done. You can then edit and run again. Your original program file is unaffected until you decide to save the edited program with

---

* Do not associate the file extension **.src** with the Desire editor as in earlier versions of Desire.

† You can drag the Graph Window to any convenient location, or position it (for the current session only) with the command-mode or programmed statement **display W x,y**.

a **keep** command or with the Editor's **Save** button. You can also save different edited versions of your program with **keep** *'filespec'* or with the Editor's **Save as** menu command.

To *pause a simulation run*, double- or triple-click the Graph Window under Windows (type **Ctrl-c** under Linux). Type **Enter** in the Command Window to *resume* the run, or **space** followed by **Enter** to *terminate the run*. Before and after a simulation, the Command Window also accepts a variety of typed commands for controlling and debugging simulations (Chapter 5, Section 5.5.2), and even operating system commands and calls to external programs.

1.3.5 Interactive Modeling with Multiple Editor Windows

Typing **ed** in the Command Window *creates a new Editor Window* containing the currently active program.* You can create as many Editor Windows as you like and

- Clear them by clicking the **New** icon to *enter additional programs*
- Use them to *preserve different edited versions of a program*
- *Drag-and-drop different user programs* into them

You can then *execute the program displayed in any one of the Editor Windows* by clicking its red **OK** button and typing **erun** or **zz** in the activated Command Window. This can be extraordinarily useful for comparing different models, or different versions of the same model, for example, with changed parameters or program lines. A typed **reload** command reloads the last-saved version of the program and closes all Editor Windows.

The typed command **keep** replaces the original user program with the edited file. **keep** *'filespec'* saves the edited program in the file specified by *filespec*.

1.3.6 A Complete Simulation Program

Properly designed software hides its advanced features until a user actually needs them. We start with the short example program in Figure 1.5. Simple though it is, this is the complete program for a small simulation study similar to the one in Figure 1.1. The simulation generates the step

---

* Typing **edit** instead of **ed** creates an Editor Window with Desire line numbers.

```
-- Step Response of Linear System
---------------------------------------------------------
display N -18 | display C16 | display R
TMAX = 0.5 | DT = 0.0001 | NN = 1000
ww = 400 | u = 440
---------------------------------------------------------
for i = 1 to 5
  r = 10 * i
  drunr
  display 2
  next
---------------------------------------------------------
DYNAMIC
---------------------------------------------------------
d/dt x = xdot
d/dt xdot = - ww * x - r * xdot + u
--
OUT | X = x – scale
dispt X
```

FIGURE 1.5   Complete program for a small simulation study.

response of a linear oscillator for different damping coefficients and produces the time-history plots in Figure 1.6.

Statements or lines starting with – (two hyphens) are *comments*, and | is a *statement delimiter*. Successive experiment-protocol-script lines in Figure 1.5 set display colors and curve thickness (Section 1.3.7), the *simulation parameters* **TMAX**, **DT**, and **NN** (Section 1.2.4),* and two *model parameters* **ww** and **u**. The initial values of the **t** and of the state variables **x** and **xdot** are not specified and automatically default to 0.

Next, *an experiment-protocol* **for** *loop* (Sec. 5.2.3) *calls five successive simulation runs* with increasing values of the damping coefficient **r**. Each **drunr** statement (equivalent to **drun | reset**) calls a simulation run and resets **t** and the differential-equation state variables to their saved initial values. The **display 2** statement keeps successive time-history curves on the same display.

The **DYNAMIC** statement indicates the start of a DYNAMIC program segment. Two derivative assignments

---

* If **NN** and **DT** are not specified, **NN** automatically defaults to the smaller of **X/DT + 1**, and 251, and **DT** defaults to **0.5 COMINT = 0.5 TMAX/(NN-1)**.

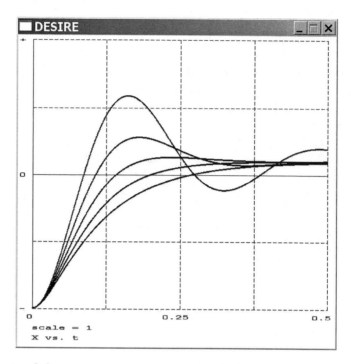

```
x = 0.8                          | -- initial value
--
drun                             | -- make a simulation run!
-----------------------------------------------------------------
DYNAMIC
-----------------------------------------------------------------
d/dt x = xdot    |    d/dt xdot = - k * x - r * xdot
--
dispt x. xdot              | -- runtime display
```

FIGURE 1.6   Graph Window produced by the simulation program in Figure 1.5.

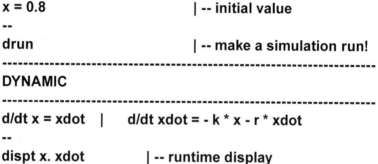

```
d/dt x = xdot
d/dt xdot = - ww * x - r * xdot + u
```

implement our two differential equations. **dispt X** plots **X = x - scale** rather than **x** for a more pleasing display (see also Section 1.3.7.2). Since **X** is needed only at output-sampling points, one can place an **OUT** statement before the defined-variable assignment **X = x - scale** (Section 1.2.3).*

---

* This time-saving **OUT** statement could be omitted in such a small program.

1.3.7 Time-History Output

1.3.7.1 Programming Time-History Graphs and Listings

Time-history graphs between **t = t0** and **t = t0 + TMAX** are produced by the DYNAMIC-segment output requests

    dispt y1, y2, ...   (up to 12 variables versus **t**)
    dispxy x, y1, y2, ...   (up to 11 variables versus the first)
    DISPXY x1, y1, x2, y2, ... (up to 6 pairs of *xy* plots)

The variables in each *display list* must be defined earlier in the program. A DYNAMIC program segment may have only one output request.

The program in Figure 1.5 generates the Graph Window shown in Figure 1.6. Our experiment-protocol line

    display N -18 | display C16 | display R

has produced a white background for publication, a black coordinate grid, and thick curves. Note that thick curves (**display R**) take some extra time. Table 1.1 lists all display commands. Graphs normally display in color. Each graph shows the current value of **scale** (see below), and *color keys* like those at the bottom of the Graph Window in Figure 1.1 relate successive curve colors to corresponding variables in the display list. Black-and-white displays like the one in Figure 1.6 have no color keys.

The output request

    type x1, x2, ...

lists the time histories of up to five variables on the display screen.

1.3.7.2 Display Scaling and Stripchart-Type Displays

The simulation parameter **scale** keeps excursions of all display-list variables **x, y1, y2, ...** between **- scale** and **scale**. **scale** defaults to 1 if no other value is specified by the experiment protocol. To *rescale individual variables*, we simply introduce new display variables such as **X = a * x**, **Y1 = b1 * y1**, **Y2 = b2 * y2**, ... , as needed. *Automatic display scaling* is discussed in Chapter 6, Section 6.8.1.

It is often useful to plot multiple time histories on different time axes. To this end, we display *offset and rescaled display variables* such as

TABLE 1.1   Display Control

Experiment-protocol scripts control runtime displays with the following programmed or command-mode statements:

display 0	Turns the simulation-run display OFF
display 1	Display ON; new runs start new displays (default)
display 2	Prevents erasing the display between runs (to combine graphs from multiple runs and/or multiple DYNAMIC program segments). Not used in command mode.
display F	Erases the Graph Window
display Q	Graphs use small dots (default)
display R	Graphs use thick dots (takes more time)
display W ix, iy	Moves the upper-left Graph Window corner to (ix, iy)
display A	Opens a Graph Window for interpreter plots (Chapter 8, Section 8.3.2)
display B	Opens an interpreter-plot Graph Window with nonnegative abscissas
display C n or display Cn	Sets the coordinate net to color **n**
display N n or display Nn	Sets the color of the first curve to color **n**. *Color keys* relate the colors of up to 8 curves to variable names in the display list.

The default background color is black. Under Windows, interpreter plots (Chapter 8, Section 8.3.2) always have a black background.

For Windows simulation-run displays, **display N n** with **n < 0** produces a white background **display N 18** and **display N -18** cause all curves to have the same color (black with **display C 16**). To remove the coordinate net (e.g., for three-dimensional plots), use **display C 16** (black) with a black background and **display C15** (white) with a white background.

Under Linux, **display C 17** produces black coordinate lines and a white background. For Linux simulation run displays, **display N 15** causes all curves to be white, and **display N 16** makes all curves black. **Display C 16** removes the coordinate net.

$$Y1 = 0.5 * (y1 + scale) \qquad Y2 = 0.5 * (y2 - scale)$$

Note that this effectively changes the vertical display scales. Figures 3.3 and 3.8 show examples.

1.3.7.3 Time-History Storage and Printing

Modified **type** statements can direct listings to text files (Chapter 5, Section 5.4.1), and **stash** statements can save time histories in binary form. But *the best way to save a simulation-run time history is with a* **store** *statement that saves the time history in an array declared by the experiment-protocol script* (Chapter 5, Section 5.3.1). Experiment protocols can file or print such time-history arrays and manipulate them, say for rescaling or fast Fourier transforms (Chapter 8, Section 8.4.1). To prepare graphs with more elaborate labels and/or scaling, you can transfer time-history arrays to programs such as Gnuplot. Time-history arrays can also produce time-function input for other DYNAMIC program segments (Chapter 6, Section 6.8.1).

1.3.8 Preparing Publication Copy

Text in Desire Editor Windows can be highlighted and copied into word processors and other programs. Graph Windows captured with **Alt-Prt Scr** under Windows or with a snapshot command under Linux can be pasted into word processors or graphics editors for storage and printing. That takes only seconds, so that you can try different versions of graphs interactively. The graphs in this book were prepared in this manner.

To create black-on-white graphs for publication use **display N - 8 | display C16** under Windows or **display N16 | display C17** under Linux. Thick curves (**display R**) may look better in print, but thin curves (**display Q**) provide better resolution and take less time.

REFERENCES

1. Korn, G.A. (2007), *Advanced Dynamic-System Simulation: Model-Replication Techniques and Monte Carlo Simulation*, Wiley, Hoboken, NJ.
2. Cellier, F., and E. Kofman (2006), *Continuous-System Simulation*, Springer, New York.
3. Cellier, F. (2010), *Numerical Simulation of Dynamic Systems*, Springer, New York.

CHAPTER 2

A Gallery of Simple Simulation Programs

2.1 INTRODUCTION

2.1.1 Basics

Properly designed software hides its advanced features until users actually need them. Thus, Sections 2.1.1 and 2.1.2 show basic rules for programming simple simulations, and Chapters 4 through 6 will add more program features.

Both experiment-protocol scripts and DYNAMIC program segments evaluate *expressions* such as*

Elevator = 3 * coeff7 - 14.4 * (sin(w * t) + Y * a$bcdE) - 1.2E-07

using double-precision floating-point arithmetic. Programs are *case-sensitive*, so that **dwarf7**, **DWARF7**, and **Dwarf7** are three different identifiers. The number π = **3.14159 ...** is designated as **PI**.

2.1.2 Experiment-Protocol Programs and Commands

The simulations in this chapter, similar to our first experiment in Chapter 1, Section 1.3.6, employ differential equation models. Each simulation

---

* Note that **2^2^3 = 2^(2^3) = 256** (as in Fortran). But there is no need to remember the precedence convention; simply use parentheses!

program has a DYNAMIC program segment defining a dynamic-system model. And each program starts with an experiment-protocol script that controls the simulation, just as a Basic program might control an automated chemistry experiment. A minimal experiment-protocol script sets the simulation parameters **TMAX**, **NN**, and **DT** defined in Chapter 1, assigns system parameters and/or initial values, and then calls a simulation run with **drun**, as in

> **TMAX = 100 | NN = 1001 | DT = 0.01**
> **alpha = - 0.555**
> **drun**

The *integration rule* defaults to **irule 1** (fixed-step 2nd-order Runge–Kutta integration), and the initial value **t0** of **t** defaults to 0. *

Most experiment-protocol statements work as *interactive commands* as well as in scripts. In particular, the command **new** erases the program currently in memory, and **ed** or **edit** displays it in a new Editor Window. **time** produces the current civil time and can also be used to time-stamp programs. The programmed or command-mode statements **reset** and **drunr** (equivalent to **drun | reset**) reset **t**, **DT**, and all differential equation state variables to the values they had at the start of the current simulation run.

Programmed **STOP** statements in experiment-protocol scripts stop execution until the user types **go**. **STOP** cannot be used in command mode, but under Linux **Ctrl-c** stops execution even during simulation runs.† Stopping a program lets you look at the display and check on the progress of a simulation study. A programmed or command-mode **write x** statement displays the current value of **x**. Section 5.5.2 (Chapter 5) will discuss additional debugging procedures.

---

* The program reminds you if you forget to specify the simulation-run time **TMAX**. The number **NN** of output points defaults to the smaller of 251 or **TMAX/DT + 1**, and the integration step **DT** defaults to ½TMAX/(NN −1). With fixed-step integration rules **DT** should be an integral fraction of the communication interval **COMINT = TMAX/(NN - 1)** (Chapter 1, Section 1.2.4.1). The Reference Manual has default parameter values for variable-step integration.

† Under Windows, execution can be paused only during simulation runs with graphics (double- or triple-click the Graph Window; Chapter 1, Section 1.3.4.2).

More elaborate experiment protocols require branching, program loops, and also file and device input/output. These topics will be discussed in Chapter 5.

2.1.3 **term** and **if** Statements in DYNAMIC Program Segments

A DYNAMIC-segment statement

term *expression*

terminates the simulation run as soon as the value of expression exceeds 0 at the end of an integration step. If the DYNAMIC segment contains more than one **term** statement, the run terminates when the first termination expression exceeds 0.

Except for input/output requests such as **dispt**, DYNAMIC-segment code subsequent to a DYNAMIC-segment **if** statement

if *scalar expression*

executes only when the expression is positive. A DYNAMIC program segment may contain multiple and/or nested **if** statements.

2.2 EXAMPLES FROM PHYSICS

2.2.1 Classical Applications and Higher-Order Differential Equations

Classical dynamic-system models in mechanics and electric-circuit problems are normally formulated in terms of second-order differential equations. These are easily reduced to the first-order differential equations (Equation 1.2a) used in our simulation programs. Any **n**th-order differential equation

$$dx^{(n-1)}/dt = F[t; x, x^{(1)}, x^{(2)}, \ldots, x^{(n-1)}] \qquad (2.1a)$$

relating a variable $x = x(t)$ and its time derivatives

$$dx/dt \equiv x^{(1)}, dx^{(1)}/dt \equiv x^{(2)}, \ldots, dx^{(n-2)}/dt \equiv x^{(n-1)} \qquad (2.1b)$$

is equivalent to a system of **n** first-order state equations with **n** state variables $x, x^{(1)}, x^{(2)}, \ldots, x^{(n-1)}$.

Classical dynamics, for example, predicts the vertical position $y = y(t)$ of a particle in free fall and no air resistance with the equation of motion

$$\textbf{d}^2\textbf{y/dt}^2 = \textbf{- g}$$

where $\textbf{g}$ = 32.2 ft/s^2 is the acceleration of gravity. This 2nd-order differential equation is equivalent to two 1st-order state equations

$$\textbf{dy/dt = ydot} \qquad \textbf{d(ydot)/dt = - g}$$

A Desire DYNAMIC program segment models these differential equations with corresponding derivative assignments

$$\textbf{d/dt y = ydot} \qquad \textbf{d/dt ydot = - g}$$

Similarly, the 2nd-order differential equation for a classical damped harmonic oscillator

$$\textbf{d}^2\textbf{x/dt}^2 = \textbf{- ww x - r dx/dt} \qquad \textbf{(2.2a)}$$

is equivalent to two 1st-order state equations

$$\textbf{dx/dt = xdot} \qquad \textbf{d(xdot)/dt = - ww} \cdot \textbf{x - r xdot} \qquad \textbf{(2.2b)}$$

modeled with the derivative assignments

$$\textbf{d/dt x = xdot} \qquad \textbf{d/dt xdot = - ww * x - r * xdot}$$

as in Chapter 1, Section 1.3.6.

2.2.2 Nonlinear Oscillators and Phase-Plane Plots
2.2.2.1 Van der Pol's Differential Equation

It is interesting to substitute nonlinear elastic and damping terms in the harmonic-oscillator model of Equation 2.2. This has useful practical applications but usually results in differential equations that cannot be solved analytically. Computer simulation permits thorough studies of such problems. In addition to simple time-history plots, computer-generated graphs of $\textbf{xdot}$ versus $\textbf{x}$ (*phase-plane plots*) can provide valuable insight into the nature of nonlinear oscillations.

As an example, the program in Figure 2.1 simulates a nonlinear oscillator defined by *Van der Pol's differential equation*. This is a 2nd-order

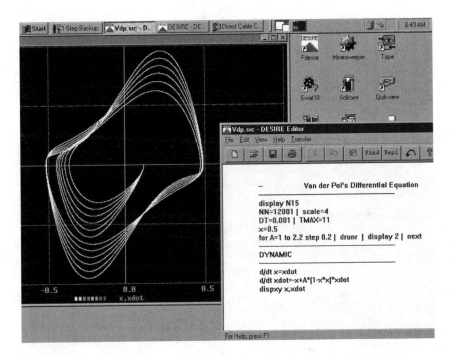

FIGURE 2.1a Windows display showing solutions of Van der Pol's differential equation for different values of the parameter $_A$.

differential equation equivalent to two state equations modeled with the derivative assignments

$$\text{d/dt x = xdot } | \text{ d/dt xdot = x - A * (x\char`^2 - 1) * xdot} \quad (2.3)$$

The nonlinear damping term **- A * (x^2 - 1) * xdot** increases the oscillation amplitude when **|x|** is small and decreases the amplitude for large values of **|x|**. All small or large initial disturbances produce the constant-amplitude *limit-cycle oscillation* illustrated in Figure 2.1. Such behavior is typical of many electronic oscillators.

Much as in Chapter 1, Section 1.3.6, the small simulation study in Figure 2.1a employs **drunr** and **display 2** statements in a DYNAMIC program segment to produce multiple phase-plane plots (**xdot** versus **x**) for successive values of the model parameter **A** on the same display. For good measure, Figure 2.1b shows time histories of **theta** and **thedot**.

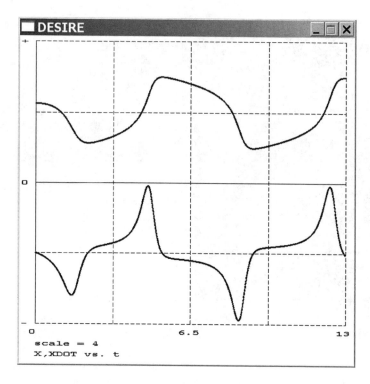

FIGURE 2.1b   Stripchart-type time-history displays for Van der Pol's differential equation are obtained by defining offset and rescaled display variables.

2.2.2.2 Simulation of a Simple Pendulum

The angular displacement **theta** of the simple rigid pendulum in Figure 2.2 satisfies the nonlinear differential equation of motion

$$\textbf{d}^2\textbf{(theta)/dt}^2 = \textbf{- ww sin(theta) - r d(theta)/dt} \qquad \textbf{(2.4a)}$$

with

$$\textbf{ww = g/Length} \quad \textbf{r = R/(mass · Length}^2\textbf{)} \quad \textbf{mass = weight/g}$$
$$\textbf{(2.4b)}$$

g is the acceleration of gravity, and **R** is a viscous-damping coefficient. For small angular displacements **sin(theta) ≈ theta**, so that the pendulum acts like the linear harmonic oscillator in Section 2.1.1. For larger values of **|theta|** we have a nonlinear oscillator.

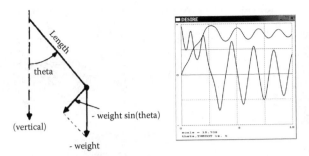

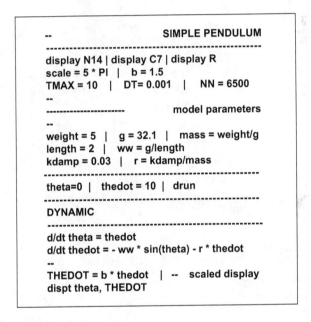

FIGURE 2.2   Pendulum geometry, program, and results.

The equation of motion (Equation 2.4) is again equivalent to two 1st-order state equations and can be modeled with

d/dt theta = thedot

d/dt thedot = - ww * sin(theta) - r * thedot

Figure 2.2 shows a simulation program and time histories of **theta** and **thedot**.

Depending on the initial displacement and angular velocity, the pendulum may "go over the top" a number of times before settling into a damped oscillation. This is illustrated by Figure 2.3, which shows the program and

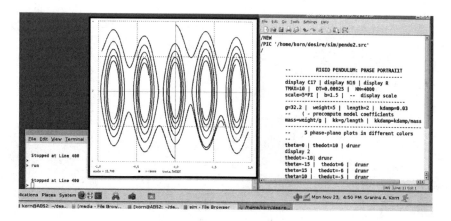

FIGURE 2.3   This Linux screen shows five pendulum phase–plane plots for different initial angular displacements and velocities. The original display showed each phase–plane plot in a different color.

results for a multirun experiment that generates five phase-plane plots for five different initial conditions on the same display. The for loop (Chapter 5, Section 5.2.3) in Figure 2.3 selects different initial values and also produces a new color for each curve.

2.2.2.3 Lorenz Differential Equations Produce Chaos

Simple simulation programs can demonstrate chaos-type behavior. *Lorenz's nonlinear differential equation system*

$$dx/dt = a * (y - x)$$
$$dy/dt = x * (b - z) - y$$
$$dz/dt = x * y - c * z$$

with, say, **a = 10, b = 28, c = 2.66667** exhibits "strange attractor" points (**x, y, z**) near which very small initial-value changes divert the solution points toward radically different solutions. Figure 2.4 displays a generalized phase-plane plot that switches "chaotically" between different modes of behavior.

2.2.3 A Simple Nuclear Reactor Simulation

The program in Figure 2.5 simulates the chain reaction in a TRIGA nuclear reactor with a simplified model [1]. The figure also shows the resulting time history of the reactor power density after a control rod is pulled out.

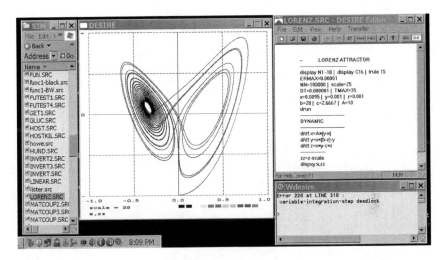

FIGURE 2.4 Simulation program and results for the Lorenz differential equations. The filemanager window lists additional available examples.

The TRIGA educational reactor is designed so that excessive temperature slows the chain reaction, and only a very short pulse results.

Our simple model neglects chain reaction "poisoning" by xenon and other reaction products. The entire reactor is "lumped" into a single-point region, so that effects of the reactor geometry on heat transfer and neutron losses are not considered. More realistic reactor and nuclear power plant models represent the reactor in terms of multiple "lumps," using possibly hundreds of state variables. One can thus model the effects of boundary losses, moderators, control-rod motion, and heat exchangers. Such simulations are important for engineering design, safety studies, and operator training.

2.2.4 An Electric Circuit Simulation with 1401 Differential Equations

Electric circuit simulation is, all by itself, so large and important a field that it is served by specialized software, such as SPICE [2] and its many derivatives. Simulation of nonlinear electric circuits (e.g., transistor circuits) normally requires a preliminary calculation of the *operating points* for all state variables, that is, of the steady-state values assumed when all devices are given their proper d-c bias voltages, and all state derivatives are zero. For large electrical circuits, the required steady-state solution of the system equations can involve a substantial amount of computation. Circuit and device models and programs are discussed in Reference 3.

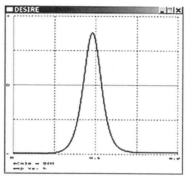

```
--   NUCLEAR – REACTOR SIMULATION
--------------------------------------------------------------------------------
--   REACTIVITY = R DOLLARS
--   NORMALIZED POWER = EN (INITIAL VALUE 1.0)
--   POWER IN MW = ENP
--   INITIAL POWER IN MW = EN0
--   NORMALIZED PRECURSOR DENSITY = Di (i = 1, 2, . . . , 6)
--   BETA/L = BOL, LAMBDA  = Ali (i = 1, 2, ..., 6)
--   NORMALIZED DELAYED NEUTRON FRACTION
        = Fi (i = 1, 2, . . . 6)
--   TEMP. COEFF. OF REACTIVITY (DOLLAR/CDGE) = ALF
--   RECIPROCAL HEAT CAPACITY (CDEG/MJ) = AK
------------------------
TMAX = 0.2   |   NN = 4001   |   scale = 200
A = 2.0   |   B = 0.0   |   BOL = 140.0
AL1 = 0.0124   |   F1 = 0.033   |   AL2 = 0.0305   |   F2 = 0.219
AL3 = 0.111   |   F3 = 0.196   |   AL4 = 0.301   |   F4 = 0.395
AL5 = 1.14   |   F5 = 0.115   |   AL6 = 3.01   |   F6 = 0.042
ALF = 0.016   |   AK = 12.5   |   GAMMA = .0267   |   EN0 = .001
EN = 1   |   D1 = 1   |   D2 = 1   |   D3 = 1   |   D4 = 1   |   D5 = 1   |   D6 = 1
----------
display N14   |   display C7   |   drun
--------------------------------------------------------------------------------
DYNAMIC
--------------------------------------------------------------------------------
R=A + B*t – ALF * TEMP
SUM = F1 * D1+F2 * D2+F3 * D3+F4 * D4+F5 * D5+F6 * D6
ENDOT = BOL * ((R - 1.0) * EN + SUM)
OMEGA = ENDOT/EN     |     ENLOG = 0.4342945 * | D(EN)
--
d/dt. EN = ENDOT
d/dt. D1 = AL1 * (EN – D1)   |   d/dt. D2 = AL2 * (EN – D2)
d/dt. D3 = AL3 * (EN – D3)   |   d/dt. D4 = AL4 * (EN – D4)
d/dt. D5 = AL5 * (EN – D5)   |   d/dt. D6 = AL6 * (EN – D6)
d/dt TEMP = AK * END * EN – GAMMA * TEMP
------------------------------
enp = EN0 * En - scale   |   dispt enp   |   --   offset display
```

FIGURE 2.5 A program for nuclear reactor simulation. The TRIGA educational reactor is designed so that any appreciable increase in neutron density reduces the reactivity. Here, an initial disturbance resulted in a short (and safe) pulse (based on Reference 1).

General-purpose differential equation solvers such as Desire can model many linear and nonlinear electric circuits quite well if you need only an occasional circuit simulation, perhaps as part of a control system study. Figures 2.6 and 2.7 illustrate the simulation of an inductance/capacitance delay-line circuit, whose state variables—inductor currents and capacitor voltages—satisfy 1401 1st-order differential equations. The computer solutions in Figure 2.7 exhibit the step-input response of the simulated electrical circuit with matched and unmatched line-termination resistances; note the signal reflection caused by the unmatched line.

We included this program here to demonstrate that a program with 1401 explicit derivative assignments works perfectly well. However, it is clearly not practical to key in such long programs. Knowledgeable

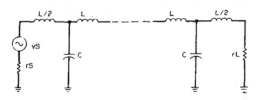

```
--    DELAY-LINE SIMULATION  (2n + 1 first-order differential equations)
---------------------------------------------------------------------------
DT = 0.0001 | TMAX =1.5 | NN = 1 + TMAX/DT | display N15 | display C7
irule 3 |  --                        4th-order Runge-Kutta integration
C = 2.0E - 04  |  L = 0.5  |  --              capacitance and inductance
a = 1/C  |  b = 1/L
rS = 50  | --                                         source resistance
rL = 50  | -- termination resistance; or L = 200 for unmatched termination
vS = 1  | --                                          step-voltage input
---
n = 700 | drun
------------------------------------------------------------------------------

DYNAMIC
------------------------------------------------------------------------------

d/dt i0 = 2 * b * (V1 + i0 * rS - vS) |  --        input current
--

d/dt V1 = a * (i0 - i1)  |  d/dt I1 = b * (V2 - V1)
d/dt V2 = a * (i1 - i2)  |  d/dt i2 = b * (V3 - V2)
. . . . . . . . . . . . . . . . . . . . . . . . . . . . .
d/dt V699 = a * (i698 - i699)  |  d/dt i699 = b * (V700 - V699)

d/dt V700 = a * (i699 - i700)
d/dt i700 = 2 * b * (i700 * rL - V700)  |  --     output current
```

FIGURE 2.6   Electric circuit delay line and an early simulation program that solved 1583 differential equations using explicit derivative assignments. The improved program in Chapter 6, Section 6.4.2, replaces all 1401 derivative assignments with just three vector operations.

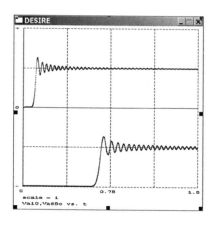 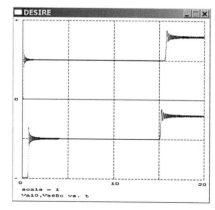

FIGURE 2.7 Time histories of the delay-line step response with matched termination resistance **rL** of 50 ohms, and with an unmatched 200 ohm resistance. Note the reflection of the input step in the unmatched case.

programmers might suggest using a program loop with subscripted variables **i[k]** and **V[k]** in the DYNAMIC segment. But it is far simpler to use Desire's vector notation, which causes no runtime loop overhead and is easy to read and write. In Chapter 6, Section 6.4.2, we shall replace 1400 of the derivative assignments in Figure 2.6 with just *two* vector assignments.

2.3 AEROSPACE AND RELATED APPLICATIONS

2.3.1 Ballistic Trajectories

Aircraft, missiles, and space vehicles are expensive to build and operate, and design decisions are needed well before any experimental hardware is built. Dynamic-system simulation is, therefore, literally indispensable for aerospace engineering and the related control system problems.

The spherical 1776-vintage cannonball in Figure 2.8a is subject to the vertical force of gravity **mg** and to aerodynamic drag opposing the instantaneous velocity direction. Since the ancient projectile is relatively slow, its drag is roughly proportional to the square of the velocity **v**. Resolving the forces along earth-fixed **x** and **y** axes and referring to Figure 2.8a, we have

$$\cos \theta = \text{xdot}/v \quad \sin \theta = \text{ydot}/v$$

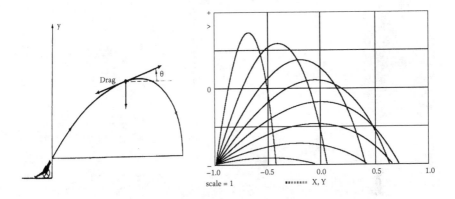

```
  --                        CANNONBALL TRAJECTORIES
------------------------------------------------------------------------------
DT = 0.05   |   TMAX = 100   |   NN=2000   |   --          timing
display N14
------------------------------------
R = 7.5E-05   |   g = 32.2
v0 = 900   |   --                                   muzzle velocity
--                                                  change elevation
for theta = 10 to 80 step 10   |   --                  in degrees
   THETA = theta * PI/180   |   --                     in radians
   xdot = v0 * cos(THETA)   |   ydot = v0 * sin(THETA)
   drunr   |   display 2   |   --                   run, don't erase
   next
display 1   |   --                                  restore display
------------------------------------------------------------------------------
DYNAMIC
------------------------------------------------------------------------------
v = sqrt(xdot * xdot + ydot * ydot)
d/dt x = xdot   |   d/dt y = ydot
d/dt xdot = - R * v * xdot   |   d/dt ydot = - R * v * ydot - g
--
term - y - 0.005   |   --              terminate each run on impact
---------------------
X = 0.00015 * x – 0.99   |   Y = 0.00025 *y – 0.99   |   dispxy X, Y
```

FIGURE 2.8A Cannonball-trajectory simulation. A loop in the experiment protocol computes the initial **xdot** and **ydot** values for different elevations. The **display 2** statement again plots successive trajectories without erasing the display. The statement **term y − 0.005** (Section 2.1.3) terminates each simulation run when the altitude **y** is almost 0.

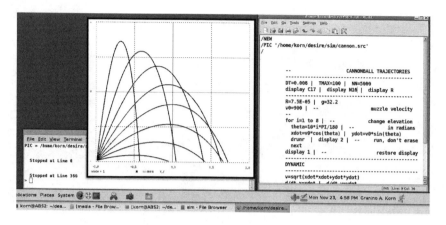

FIGURE 2.8b   Linux display of the cannonball simulation.

so that the equations of motion are

$$(d/dt)\ x = xdot \quad (d/dt)\ xdot = -Rv^2 \cos\theta = -Rv \cdot xdot$$
$$(d/dt)\ y = ydot \quad (d/dt)\ ydot = -Rv^2 \sin\theta - g = -Rv \cdot ydot - g$$

with

$$v = sqrt(xdot^2 + ydot^2)$$

Using engineering units, **g = 32.2** is the acceleration of gravity, and **R = 7.5E-05** is a drag coefficient divided by the projectile mass.

In the simulation program of Figure 2.8a, the experiment protocol employs the loop

```
for theta = 10 to 80 step 10
    THETA = theta * PI /180
    xdot = v0 * cos(THETA) | ydot = v0 * sin(THETA)
    drunr | display 2
    next
```

to precompute initial values of **xdot** and **ydot** for successive simulation runs with increasing gun-elevation angles **theta**. The loop also changes the number of degrees **theta** to the corresponding value **THETA** in radians. Figure 2.8b shows a Linux display.

2.3.2 Simple Flight Simulation

2.3.2.1 Pitch-Plane Flight Equations

Reference 4 reviews aircraft flight dynamics. To avoid unnecessary complication, we will discuss *two-dimensional* aircraft motion in the longitudinal or pitch plane. Referring to Figure 2.9, we resolve all accelerations along *wind axes* along and perpendicular to the *velocity vector*. We use six state variables:

The coordinates **x** and **y** of the aircraft center of gravity
The *aircraft velocity* **v**
The *flight-path angle* **theta1** between the velocity vector and the horizontal **x** axis
The *pitch angle* **phi1**
The *pitch angular velocity* **phi1dot**

The acceleration components are **dv/dt** in the velocity direction and **v d(theta1)/dt** at right angles to the instantaneous velocity, so that we have

$$\textbf{(d/dt) v = - mg sin(theta1) + thrust} \cdot \textbf{cos(theta1) - drag}$$
$$\textbf{(d/dt) theta1 = [- mg cos(theta1) + thrust} \cdot \textbf{sin(theta1) + lift]/mv}$$
$$\textbf{(2.5a)}$$

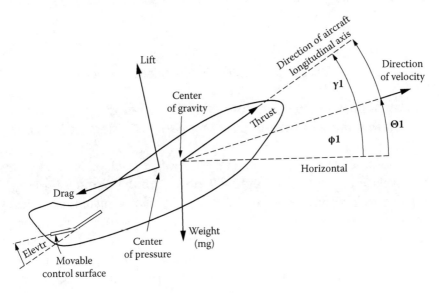

FIGURE 2.9   Aircraft motion in the pitch plane.

We add the rotational equations of motion

$$\text{(d/dt) phi 1= phi1dot} \quad \text{(d/dt) phi1dot = M/I} \qquad \text{(2.5b)}$$

where **M** is the aerodynamic moment, and $_I$ is the aircraft moment of inertia about its pitch axis. **Drag, lift**, and **M** are functions of the aircraft velocity **v**, of the *angle of attack*

$$\text{gamma1 = phi1 - theta1} \qquad \text{(2.5c)}$$

and also of the aircraft altitude (which determines the air density), of the elevator deflection **elevtr**, and of various engine control inputs such as throttle settings. The aerodynamic functions are usually obtained from stored wind-tunnel measurements. Interpolation in the resulting large multiargument function tables (there are even more arguments in the three-dimensional case) is a major part of the computer load in realistic flight simulations.

The four equations (2.5) are the *pitch-plane flight* equations. Computation of the aircraft trajectory with respect to earth-fixed **x** and **y** axes (*earth axes*) requires two additional state equations:

$$\text{dx/dt = v cos(theta1)} \quad \text{dy/dt = v sin(theta1)} \qquad \text{(2.6)}$$

2.3.2.2 Linearized Flight Equations

It is sometimes sufficient to consider only small perturbations

$$\text{phi = phi1 - phi0} \quad \text{phidot = phi1dot - phidot0}$$
$$\text{theta = theta1 - theta0} \quad \text{gamma = gamma1 - gamma0} \qquad \text{(2.7)}$$

of some steady-state flight condition, say, straight and level flight at constant velocity **v** = **v0**, with **theta1 = 0**, **phi1 = phi0 = gamma0**, and all state-variable derivatives equal to 0. This is of practical interest for studies of aircraft stability and autopilot design, where flight-path perturbations are *supposed* to remain small. We substitute the expressions (2.7) into the nonlinear flight equations (Equation 2.5) and replace their right-hand sides by Taylor-series expansions about the steady-state conditions. We keep only terms linear in the small perturbations (2.7) and obtain the *linearized flight equations*

$$\text{(d/dt) theta} = (\text{L1} \cdot \text{gamma} + \text{L2} \cdot \text{elevtr}) / (m \cdot v0)$$
$$\text{(d/dt) phi} = \text{phidot}$$
$$\text{(d/dt) phidot} = (\text{M1} \cdot \text{gamma} + \text{M2} \cdot \text{phidot} + \text{M3} \cdot \text{elevtr})/l$$
$$\text{gamma} = \text{phi} - \text{theta} \tag{2.8}$$

where **L1, L2, M1, M2, M3** are *aerodynamic derivatives* obtained from wind-tunnel data.

Figure 2.10 illustrates a linearized flight simulation for a propeller-driven aircraft. This simple linear model exhibits the typical time delays associated with pitch and altitude control. Specifically, elevator deflections must change the pitch angle before the resulting increased angle of attack increases the flight-path angle and thus lifts the aircraft. **M2** is negative, so that the pitch rate causes a damping torque (aerodynamic damping). The simulated pitch response to elevator-deflection steps in Figure 2.10 exhibits the well-known damped "phugoid" oscillation.

2.3.3 A Simplified Autopilot

A basic automatic pilot produces control-surface deflections that minimize attitude errors sensed by an *attitude gyro*. In Figure 2.11, a pitch-stabilizing autopilot controls the elevator deflection **elevtr** as a function of the gyro-measured pitch-angle perturbation **phi**. For better control, a voltage proportional to the pitch-rate perturbation **phidot** is sensed by a *rate gyro* and added to the attitude signal. The autopilot control voltage (controller output) is then

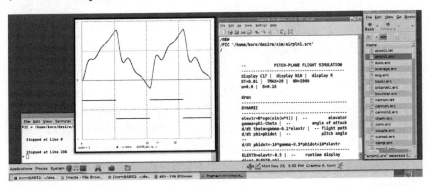

FIGURE 2.10 Linearized flight simulation in the pitch plane. This Linux display shows the complete program and the pitch-angle response to periodic positive and negative elevator step inputs.

u = - gain · phi – damp · phidot

Actually neither the gyros nor the control-surface actuators can act instantaneously, so that their delayed action must be represented by additional state equations. Instead of simulating detailed gyro and actuator equations of motion, the simple autopilot-simulation program in Figure 2.11a lumps their delays into a single extra state equation,

(d/dt) elevtr = b(a · u - elevtr)

which act like a first-order lag filter with a time constant equal to **1/b** seconds. A more realistic autopilot model will be described in Chapter 4, Sections 4.6.1 and 4.6.2.

In the program of Figure 2.11a, our simplified autopilot model controls the linearized aircraft model of Figure 2.10 to force straight and level flight after an initial pitch error. Specifically, the pitch angle **phi** is given a non-zero initial value, and the resulting pitch error is then reduced to zero by the autopilot feedback. The program lets you vary the autopilot damping coefficient **damp** to investigate the effects of the pitch-rate feedback. Figure 2.11b shows typical results. As with many control systems, more damping slows the autopilot response but damps oscillations.

Autopilot development normally proceeds from design and simulation to *partial system tests* of real autopilot components, with attitude and rate gyros placed on a *flight table* whose motion is driven by the attitude outputs of a *real-time flight simulation*, that is, a computer simulation synchronized with real time by clock interrupts.

2.3.4 Torpedo Trajectory

In Section 2.3.2, we used linearized flight equations to model small deviations from straight and level flight, but studies of maneuvering vehicles require nonlinear models. We will simulate the two-dimensional (yaw-plane) motion of a program-steered torpedo through water (Figures 2.12 and 2.13). Unlike our aircraft model, the torpedo model employs *body-axis* velocity components **u, v** along and perpendicular to the torpedo longitudinal axis as state variables. **psi** is the *yaw-attitude angle*, and **psidot** is the *yaw rate*. Referring to Figure 2.12, the *drag, side force,* and *moment* and the hydrodynamic force on the torpedo rudder are all approximately proportional to the square of **u**. We thus find longitudinal and transverse equations of motion

$$\text{(d/dt) u = (thrust - drag)/mass = UT - X1} \cdot u^2$$
$$\text{(d/dt) v = Y1} \cdot u2\sin(γ^2) + Y^2 \cdot u \cdot \text{psidot} + Y^3 \cdot u^2 \text{ rudder}$$

where **γ2** is the *yaw angle of attack* between the longitudinal axis and the velocity vector. The yaw rotation is determined by state equations similar to those for pitch in Section 2.3.1,

$$\text{(d/dt) psi = psidot}$$
$$\text{(d/dt) psidot = N1} \cdot u2 \cdot \sin(γ2) + N2 \cdot u \cdot \text{psidot} + N3 \cdot u2 \text{ rudder}$$

The **u · psidot** terms approximate hydrodynamic damping, and **UT, Y1, Y2, N1, N2, N3** are constant torpedo design parameters. Two additional state equations integrate the velocity components into trajectory coordinates **x** and **y**:

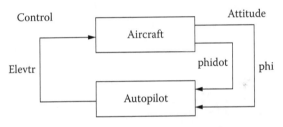

```
    --                          SIMPLE AUTOPILOT SIMULATION
- - - - - - - - - - - - - - - - - - - - - - - - - - - - - - - - - - - - - - -
DT = 0.0002 |   TMAX = 3  |   NN = 2000
display N15 | display C7 | display R
gain = 3 |  damp = 0.75 |  a = 1  |  b = 7  | --      set parameters
phi = 0.9999 |  --                             initial pitch error
drun
- - - - - - - - - - - - - - - - - - - - - - - - - - - - - - - - - - - - - - -
DYNAMIC
- - - - - - - - - - - - - - - - - - - - - - - - - - - - - - - - - - - - - - -
    --                                              AUTOPILOT
u = – gain * phi – damp * phidot  | --        autopilot controller
d/dt elevtr = b * (a * u – elevtr) | --       simplified servo model
- - - - - - - - - - - - - - - - - - - - - - - - - - AIRCRAFT FLIGHT EQUATIONS
    --
gamma = phi – theta
d/dt theta = gamma – 0.1 * elevtr
d/dt phi = phidot | d/dt phidot = –10 * gamma – 0.5 * phidot + 10 * elevtr
- - - - - - - - - - - - - - - - - - - - - - - - - - - - - - - - - - - - - - -
ELEVTR = 0.1 * elevtr – 0.5 |   dispt ELEVTR,phi  | -- runtime display
```

FIGURE 2.11a   Simulation of a simple automatic pilot system. Attitude and rate gyros sense the pitch angle **phi** and the pitch rate **phidot**, and a servomechanism positions the elevator surface. The aircraft is modeled with the linearized flight equations of Figure 2.10.

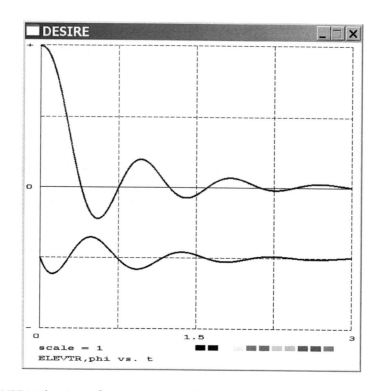

FIGURE 2.11b   Autopilot response to an initial pitch error.

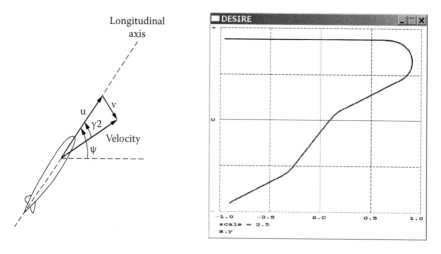

FIGURE 2.12   Torpedo geometry, and a typical program-controlled trajectory.

$$dx/dt = u \cos(psi) - v \sin(psi) \mid dy/dt = u \sin(psi) + v \cos(psi)$$

The torpedo's weathercock stability keeps the yaw angle of attack **γ2** so small that we can approximate

$$\sin(\gamma 2) \approx \tan(\gamma 2) \approx v/u$$

and obtain simplified derivative assignments

```
d/dt u = UT - X2 * u^2
d/dt v = u * (Y1 * v + Y2 * psidot + Y3 * u * rudder)
d/dt psidot = u * (N1 * v + N2 * psidot + N3 * u * rudder)
d/dt psi=psidot
```

The rudder deflection is programmed as a function of time with

rudder = func1(t; RT)

using a table-lookup/interpolation function generator (Chapter 4, Section 4.2.2). Figure 2.13 shows the complete simulation program producing the programmed turn, snake search, and attack trajectory in Figure 2.12.

2.3.5 Translunar Satellite Orbit

The satellite simulation of Figure 2.14 assumes a fixed earth exerting a simple inverse-square-law gravitational force on the satellite, and no air resistance. With the polar coordinates indicated in Figure 2.14, the state equations are modeled by the derivative assignments [4]

```
d/dt r = rdot  |  d/dt rdot = - GK/(r^2) + r * thdot^2
d/dt theta = thdot  |  d/dt thdot = 2 * rdot * thdot/r
```

Two additional defined-variable assignments,

$$x = r * \cos(theta) \mid y = r * \sin(theta)$$

produce the orbit display. The given initial orbit-injection velocity **v** is assumed to be horizontal and produces the initial value **thdot(0) = v/r(0)**. As in Figure 2.13, we used a variable-step integration routine; Figure A.3 in the Appendix shows the actual time history of the integration step size **DT**.

-- 2-DIMENSIONAL TORPEDO SIMULATION

```
display N14 | display C7 | display R | scale=2.5
DT = 0.001 | TMAX = 48 | NN = 1500
irule 15 | ERMAX = 0.1
```

```
ARRAY  RT[48] |  -- RUDDER-COMMAND  BREAKPOINT
TABLES
data 19,20,25,26 | -- circle
data 32,33,34,40,41,42,47,48,49 | -- snake
data 55,56,57,63,64,65,71,72,73
data 74,75 | -- attack
```

```
data 0,0.25,0.25,0 | -- circle
data 0,-0.25,0,0,0.25,0,0,-0.25,0,0,-0.25 | -- snake
data 0,0,-0.25,0,0,0.25,0
data 0,-0.125 | -- attack
read RT
```

```
-- TORPEDO PARAMETERS
UC = 0.25
X1= 0.8155 | X2 = 0.8155 | -- u-equation parameters
UT = X1*UC*UC
Y1 = - 15.701 | Y2 = - 0.23229 | Y3 = 2.0002 | -- v-equation
parameters
N1 = - 303.801 | N2 = - 44.866 | N3 = - 243.866 | -- r-equation
parameters
x = - 2.4 | y = 2.2
```

```
drun
```

```
DYNAMIC
```

```
rudder = func1(t; RT) | -- rudder command function
rr = rudder * u
d/dt u = UT - X2 *u^2
d/dt v = u * (Y1 * v + Y2 * psidot + Y3 * rr)
d/dt psidot = u * (N1 * v + N2 * psidot + N3 * rr)
d/dt psi = psidot
cosp = cos(psi) | sinp = sin(psi)
d/dt x = u*cosp – v * sinp | d/dt y = u * sinp + v * cosp
```

```
dispxy x, y
```

FIGURE 2.13   Complete program for the torpedo simulation.

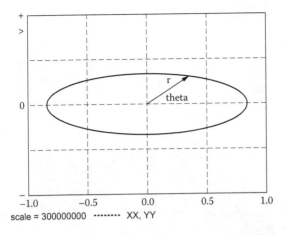

scale = 300000000 ········ XX, YY

```
--                                          SATELLITE ORBIT
- - - - - - - - - - - - - - - - - - - - - - - - - - - - - - - - - - - - - - - -
display N14  |  display C7  |  display R
NN = 3000  |  DT = 0.03  |  TMAX = 220000
scale = 3E+08  |  display N14  |  -- display scale and color
irule 15  |  ERMAX = 0.000001  |  -- Gear integration rule
- - - - - - - - - - - - - - - - - - - - - - - - - - - - - - - - - - - - - - - -
GK = 1.42E+16  |  --                            gravity constant
r = 21.0E+06  |  v = 36000  |  thdot = v/r  | -- initial values
drun
- - - - - - - - - - - - - - - - - - - - - - - - - - - - - - - - - - - - - - - -
DYNAMIC
- - - - - - - - - - - - - - - - - - - - - - - - - - - - - - - - - - - - - - - -
d/dt r = rdot  |  d/dt theta = thdot
d/dt rdot = –GK/(r^2) + r* thdot^2
d/dt thdot = –2 * rdot * thdot/r
- - - - - - - - - - - - - - - - - - - - - - - - - - - - - - - - - - - - - - - -
XX = r * cos(theta) + 0.8 * scale  |  YY = r * sin(theta)
dispxy XX, YY
```

FIGURE 2.14   Satellite orbit simulation program and result.

2.4 MODELING POPULATION DYNAMICS

2.4.1 Simulation of Epidemic Propagation

The epidemic-propagation program in Figure 2.15 exercises a simple population-dynamics model. We are interested in the population counts of susceptible, sick, and cured individuals. Although population counts are really integers, we model them as continuous state variables. Their rates of change define three state-equation assignments

d/dt suscept = - A * suscept * sick
d/dt sick = (A * suscept - B - C) * sick
d/dt cured = B * sick

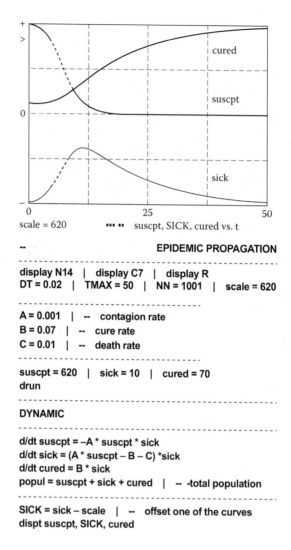

```
                        EPIDEMIC PROPAGATION
--  ---------------------------------------------------------
display N14   |   display C7   |   display R
DT = 0.02   |   TMAX = 50   |   NN = 1001   |   scale = 620
------------------------------------------------------------
A = 0.001   |   --   contagion rate
B = 0.07    |   --   cure rate
C = 0.01    |   --   death rate
------------------------------------------------------------
suscpt = 620   |   sick = 10   |   cured = 70
drun
------------------------------------------------------------

DYNAMIC
------------------------------------------------------------
d/dt suscpt = –A * suscpt * sick
d/dt sick = (A * suscpt – B – C) *sick
d/dt cured = B * sick
popul = suscpt + sick + cured   |   -- -total population
------------------------------------------------------------
SICK = sick – scale   |   --   offset one of the curves
dispt suscpt, SICK, cured
```

FIGURE 2.15 Epidemic propagation, a typical population-dynamics problem. Note that population counts are treated as continuous variables.

where **A** is the contagion rate, **B** is the cure rate, and **C** is the death rate. A cure is assumed to confer immunity. The total current population count is

popul = suscept + sick + cured

(initially nonsusceptible individuals are not considered). The solution time histories in Figure 2.15 show how the number of sick individuals first increases rapidly and then peaks and decreases to zero.

2.4.2 Simulation in Ecology: A Host–Parasite Problem

2.4.2.1 A Host–Parasite Problem

Models similar to the epidemic-propagation model describe the dynamics of two populations that destroy and/or feed one another (host-parasite or predator–prey problems, combat simulation) [5]. The program in Figure 2.16 simulates a parasite population of size **parasite** and a host population of size **host**, which is the parasites' food supply. **k1** is the difference between the host population's natural birth and death rates, and **k2** is the difference between the parasite population's natural death and birth rates. The hosts suffer an additional death rate **k4 · parasite** due to the parasite predation, which also gives the parasites an additional birth rate **k3 · host**. The resulting state equations are the *Volterra–Lotka differential equations*

$$\text{(d/dt) host} = (k1 - k4 \cdot \text{parasite}) \text{ host}$$
$$\text{(d/dt) parasite} = (k3 \cdot \text{host} - k2) \text{ parasite}$$

Figure 2.16a shows a simulation program. The resulting Windows display in Figure 2.16b illustrates how the parasite population initially grows with the host population but reduces the latter and thus its own food supply. The parasite population then declines; as a result, the host population tends to increase again. For the values of **k1**, **k2**, **k3**, **k4** used in our program, the resulting predator–prey population cycle repeats periodically, as is often observed in nature. With different parameter values (e.g., due to simulated diseases or overhunting), solutions of the Volterra–Lotka equations agree with the observed fact that both prey and predators can die out.

2.4.2.2 Generalizations

Population-dynamics models have been greatly elaborated to account for

Multiple interacting populations of plants, animals, soils, microorganisms, etc., plus masses of chemicals such as oxygen and carbon dioxide, all affecting the changes of the different state variables ("biome" models, with possibly hundreds of state variables)

Subdivision of populations into local areas with different conditions, and/or into age "cohorts" with different birth, fertility, and death rates

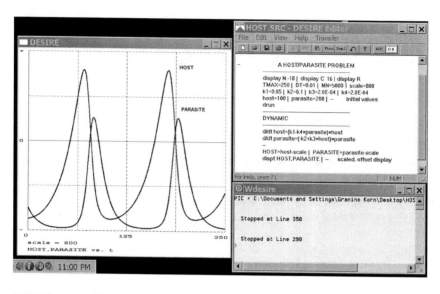

FIGURE 2.16a   Host–parasite simulation representing the dynamics of parasites feeding on a host population. Time units could be hours, days, or months.

-- A HOST-PARASITE PROBLEM
----

display N14 | display C7 | display R
TMAX = 250 | DT = 0.01 | NN = 5000 | scale = 800
k1 = 0.05 | k2 = 0.1 | k3 = 2.0E-04 | k4 = 2.0E-04
host = 100 | parasite = 200 | -- initial values
drun

DYNAMIC
----

d/dt host = (k1 - k4 * parasite) * host
d/dt parasite = (k3 * host - k2) * parasite

HOST = host – scale | PARASITE = parasite - scale
dispt HOST, PARASITE | -- scaled, offset display

FIGURE 2.16b   A Windows display of the host–parasite program. The original color-coded display clearly identified the curve with the larger excursions as the host–population curve. The parasite population grows after the host population grows, then chokes it off, and decreases from lack of food. With the parameter values used here, the populations recover, and the process repeats periodically. With different rate constants, both population could die off.

More complicated functional dependence of birth and death rates on various system variables and on the time (diurnal or annual variations of food supplies, growth rates, breeding, etc.)

Effects of fertilizers, pesticides, and pollutants

While impressive and plausibly realistic models have been developed, measurement of plausible coefficients and function values for ecological models can be very difficult. It is not easy to measure the rate at which 2-year-old coyotes eat 4-month-old rabbits. But simulations can, if nothing else, study the *sensitivity* of results to changes in parameters about which we have little information.

Models essentially similar to population-dynamics models are also very useful for studying *chemical reactions* [5,6].

REFERENCES

1. Hetrick, D. (1971), *Dynamics of Nuclear Reactors*, University of Chicago Press.
2. Thorpe, T.W. (1992), *Circuit Analysis with SPICE*, Wiley, Hoboken, NJ.
3. Howe, R.M. (1968), in Karplus, W.J., and G.A. Bekey, *Hybrid Computation*, Wiley, Hoboken, NJ.
4. Montenbruck, O., and E. Gill (2000), *Satellite Orbits*, Springer, New York.
5. Cellier, F.E. (1994), *Continuous-System Modeling*, Springer, New York.
6. Upadhayay, S.K. (2000), *Chemical Kinetics and Reaction Dynamics*, Springer, New York.

CHAPTER 3

Introduction to Control System Simulation

3.1 SIMULATION AND CONTROL SYSTEM DESIGN

3.1.1 Introduction

Control system design was the earliest successful application of computer simulation. We have already introduced a simple autopilot model in Chapter 2, Section 2.3.3. This chapter introduces basic control system modeling; additional related programs will be found in Chapters 4 and 6. Section 3.4 points the way to more elaborate control system modeling.

3.1.2 Simulation of a Simple Servomechanism

The motor of a feedback position-control system or *servomechanism* drives a load so that its output displacement **x = x(t)** follows an input **u = u(t)** after an initial transient. The servo controller generates the control voltage **V** as a function of the servo *error* **error = x − u** and the output-displacement rate **xdot = dx/dt; xdot** is measured by a tachometer connected to the output. The amplifier output voltage **V** produces the saturation-limited motor torque **torque**.

The simulation program in Figure 3.1 displays the servo response to a suddenly applied sinusoidal input **u = A cos(w t)**. For **w = 0, u(t)** becomes a *step input*. Servo controllers can be nonlinear function generators, but we will use a simple linear controller modeled with

```
--                                    A SIMPLE SERVO SIMULATION
-----------------------------------------------------------------------
scale = 2 | display N14 | display C7
TMAX = 11 | NN = 4000 | DT = 0.0005 | -- timing
-----------------------------------------------------------------------
A = 0.8 | g2 = 2 | maxtrq = 0.85 | gain = 200
w = 0.5 | r = 26 | R = 0.6
drun
-----------------------------------------------------------------------
DYNAMIC
-----------------------------------------------------------------------
u = A * cos(w * t) | error = x - u | --           input and error
V = - gain * error - r * xdot | --                controller output
torque = maxtrq * tanh(g2 * V/maxtrq) | --        dynamics
d/dt x = xdot | d/dt xdot = torque - R * xdot
-------------------------------------------------------------  display
X = x + 0.5 * scale | U = u + 0.5 * scale
TORQUE = torque-0.5 * scale | --                  offset curves
ERROR = error + 0.5 * scale
dispt X, U, TORQUE, ERROR
```

FIGURE 3.1 Complete simulation program for a simple servomechanism.

$$\text{error = x - u } V = \text{ - gain * error - r * xdot}$$
$$V = \text{ - gain * error - r * xdot} \tag{3.1a}$$

The servo gain **gain** and the damping coefficient **r** are positive controller parameters. The dynamics of motor, gears, and load are represented by

$$dx/dt = xdot \qquad d/dt\ xdot = (torque - R * xdot)/I \tag{3.1b}$$

where **I** represents the combined inertia of motor, gears, and load, and **R** is a positive motor-damping coefficient. For simplicity, we assume that **torque** and **R** are scaled so that **I = 1**. We model motor field saturation with the output-limiting hyperbolic-tangent function, so that

$$torque = maxtrq * tanh(g2 * V/maxtrq)$$
$$torque = maxtrq * sat(g2 * V/maxtrq) \tag{3.1c}$$

Figure 3.2 shows the response of the simulated servomechanism to sinusoidal and step inputs. Assuming that **maxtrq, R,** and **g2** are given

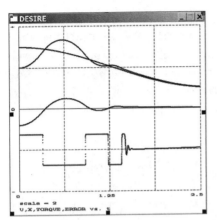
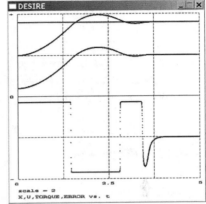

FIGURE 3.2   Cosine-function and step-function responses of the simulated servomechanism. Both displays show time histories of the servo input, output, error, and torque. The original displays were in color.

motor and amplifier parameters, we can vary the controller parameters **gain** (servo gain) and **r** (servo damping) to study their effect on the servo performance. As is well known, high gain and/or low damping produce fast servo response but cause overshoot or even oscillations and instability. More damping slows the response.

3.1.3 Simulation Studies and Parameter Optimization

3.1.3.1 Test Inputs and Error Measures

Simulation lets the control system designer change controller parameters and motor characteristics and observe the resulting behavior of servo outvput, error, and torque. The problem is to find a servo design that will, in some sense, produce desirably small errors. It will be necessary to specify a *test input* **u(t)** deemed typical for the intended application. Frequently used test inputs are steps, ramps, sinusoids, and noise. More often than not, one needs to try several different test inputs.

Simulation experiments usually begin with a seaman's-eye look at the solution time histories for the selected type of input. One tries to vary design parameters until the errors look reasonably small. Such simple-minded computer-aided experimentation gives the designer some feeling for a proposed new system. The simulation may show up control system instability or gross design errors. One also hopes to relate observations and theoretical analysis.

Objective *error measures* are functionals that depend on the entire time history of the control system error **error = x − u** for each specified input **u(t)**. Simply reading the *maximum absolute error* or the *maximum squared error* from a time-history graph is one possibility. More commonly used error measures are integrals over the error time history. One computes these simply as *extra state variables* with zero initial values:

(d/dt) IAE = |x − u| (IAE, integral absolute error)

(d/dt) ISE = (x - u)²   (ISE, integral squared error)

(d/dt) MSE = (x - u)²/TMAX   (MSE, mean squared error)

3.1.3.2 Parameter-Influence Studies

We next proceed to modify the control system model until selected error measures meet acceptable limits for the specified class of inputs, or until an error measure is as small as possible. Parameter-influence studies program multiple simulation runs to explore the effects of changing parameters. Desire automates such studies with more or less elaborate experiment-protocol scripts.

We can, for example, replace the experiment-protocol lines at the beginning of the program in Figure 3.1 with the new script

```
TMAX = 11 | NN = 4000 | DT = 0.005 | -- timing
scale = 2 | display C7
----------------------------------------------------------------
A = 0.8 | g2 = 2 | maxtrq = 0.85 | gain = 200
w = 0.5 | R = 0.6
----------------------------------------------------------------
r0 = r | --                                      base case
delta = 0.25
color = 12 | display N 12 | drunr
--
display 2 | -- show more curves on the same display
--
r = r0 * (1 + delta) | display N 11 | drunr
r = r0 * (1 - delta)  | display N 13 | drun
```

With the same DYNAMIC program segment as in Figure 3.1, our revised experiment protocol displays solution time histories **x(t)** for a *base case* (**r = r0**) and two *perturbed cases* (**r = r0 + delta** and **r = r0 − delta**) on the same display, in three different colors. Figure 3.3 shows the three step responses.

The Desire model replication technique developed in Reference 1 permits much more sophisticated parameter-influence studies. We shall present a simple example in Chapter 6, Section 6.7.1.4.

3.1.3.3 Iterative Parameter Optimization
Figures 3.4 and 3.5 demonstrate a program for *iterative parameter optimization*. The DYNAMIC program segment in Figure 3.5 models the same servomechanism we studied in Sections 3.1.2 and 3.1.3.2. An extra derivative assignment

d/dt ISE = error * error
d/dt ISE = error^2

with **ISE(0)** defaulting to 0 produces the integral squared error **ISE = ISE(TMAX)** at the end of each simulation run. The experiment protocol *automatically* adjusts the controller damping parameter **r** so as to minimize

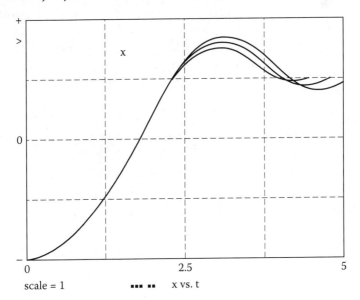

FIGURE 3.3 Results of a parameter-perturbation experiment varying the controller damping coefficient **r** by ±30%.

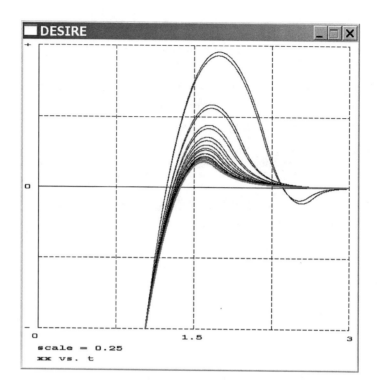

FIGURE 3.4 Automatic optimization of the servo step response. A **display 2** statement (Chapter 1, Section 1.3.6) permitted us to plot results from successive optimization runs on the same display. The graphs show how each working run is followed by a gradient-measuring run, and then by a larger working step. The original display was in color, with the optimal solution curve lighting up in red.

ISE for the specified step input **u(t)**.* The program in Figure 3.5 uses the **repeat** and **if/else/proceed** statements described in Chapter 5, Section 5.2.3.

At each stage of the optimization, a *gradient-measuring simulation run* determines the change **DISE** in **ISE** produced by a fixed trial perturbation **Dr** of the servo damping parameter **r**. A *working run* then changes **r** by a larger step **opgain * DISE** in the negative gradient direction. The optimization process is stopped when |**DISE**| becomes smaller than a specified small number **crit**. The solution display in Figure 3.4 clearly shows successive pairs of gradient-measuring and working runs.

---

* Since our torque-limited servo is not a linear system,, the optimal value or **r** will depend on the step amplitude.

```
--                    SIMPLE SERVO OPTIMIZATION
--          intentionally slowed for a better demonstration
-------------------------------------------------------------------
TMAX = 3  |  NN = 4000  |  DT = 0.0005  |  -- timing
scale = 0.25  |  display N14  |  display C7
-------------------------------------------------------------------
A = 0.8  |  g2 = 2 | maxtrq = 0.85  |  gain = 200
r = 26  |  R = 0.6 | u = 0.5
opgain= 20000 | crit = 0.00002  |  Dr = 0.8  |  -- optimizer
drun | --                             initial trial run
display 2 | --                        multiple runs on one plot
repeat
  oldISE = ISE
  r = r+Dr
  reset | drun | --              run with perturbation
  Derr = ISE - oldISE  | --      gradient measurement
  if abs(Derr) < crit then exit | --        we are done!
  else proceed
  r = r - Dr – opgain * Derr | --           working step
  reset | drun
  until 0 > 1 | --                   keep trying!
-------------------------------------------------------------------
reset | display R | display N12 | drun | -- change color
write 'optimal values: r=';r,'ISE=';ISE
-------------------------------------------------------------------
DYNAMIC
-------------------------------------------------------------------
error = x - u | --                 controller input and
V = - gain * error – r * xdot | --            output
torque = maxtrq * tanh(g2 * V/maxtrq) | --   dynamics
d/dt x = xdot | d/dt xdot = torque – R * xdot
--
d/dt ISE = error^2 | --            integral square error
-------------------------------------------------------------------
xx = 2 * x – 1 | dispt xx | --     offset runtime display
```

FIGURE 3.5 Complete program for the parameter optimization experiment. A gradient-measuring algorithm perturbs the damping coefficient **r** to minimize the integral squared error **ISE** for a unit-step input **u**. Values of the optimizer parameters **Dr** and **opgain** were deliberately selected to slow the optimization for a better demonstration.

d/dt V1 = - a * (V1 + error)| V = - gain * error + b * V1

3.1.4 Where Do We Go from Here?

3.1.4.1 More Elaborate Controllers

Our introductory servo model is deliberately simple. The linear-controller definition (Equation 3.1a) implements only *proportional error feedback* and *output-rate feedback* from a tachometer. One can easily add *integral control* by generating the time integral **y** of the servo error with

d/dt y = error

and adding the term **alpha * y** to the controller voltage **V** in Section 3.1.2.

Error-rate feedback (as opposed to the simpler output-rate feedback in Section 3.1.2) normally computes a low-pass-filtered approximation to **d(error)/dt** with a linear network simulated as in Section 3.1.4.3.

Nonlinear controllers involve function generators, including limiters, switches, and comparators. They can, for instance, implement relay servos (bang-bang control, Chapter 4, Sections 4.6.1 and 4.6.2). Desire models of *fuzzy-logic control* are treated in Reference 1.

3.1.4.2 More Elaborate Plant Models and Control System Noise

Control system design usually involves much more complicated plants than the simple servo in Section 3.1.2, in particular aerospace vehicles (Chapter 2, Sections 2.3.2 through 2.3.5) and guided missiles. Control systems must also deal with random parameter tolerances and noise. Desire programs for Monte Carlo simulation of noisy control systems are discussed in a separate textbook [1].

3.1.4.3 Control System Transfer Functions and Frequency Response

Linear controller networks, and indeed entire linear control systems, are often specified by *transfer-function input/output relations* [1]

$$\text{OUTPUT(s)} = \frac{bb\, s^n + b_n\, s^{n-1} + \ldots + b_2\, s + b_1}{s^n + a_n\, s^{n-1} + \ldots + a_2\, s + a_1} \text{INPUT(s)}$$

where **OUTPUT(s)** and **INPUT(s)** are the Laplace transforms of the system output **output(t)** and the system input **input(t)**. Replacing **s** with **jω** yields the complex *frequency-response function*, which compactly describes the response of the linear system to sinusoidal inputs of circular frequency **ω**. The transfer-function model is equivalent to an **n**th-order state-equation system

> **output = x_n + bb * input**
> **d/dt x_1 = b_1 * input − a_1 * output**
> **d/dt x_2 = x_1+ b_2 * input − a_2 * output**
>
> **. . . . .**
>
> **d/dt x_n = x_{n-1} + b_n * input − a_n * output**

We shall reduce these **n + 1** *assignments to just two simple program lines* but must postpone this until we introduce programs with subscripted variables, vectors, and matrices in Chapter 6. Chapter 6 will also demonstrate the use of fast Fourier transforms for computing the frequency response of linear systems.

DESIRE experiment-protocol scripts (but not DYNAMIC program segments) can manipulate and plot complex numbers. We can, therefore, produce Nyquist, Bode, and root-locus plots for linear control systems (Chapter 8, Section 8.3.4).

3.2 DEALING WITH SAMPLED DATA

3.2.1 Models Using Difference Equations

In control system simulations, differential equation systems variables represent "analog" variables that change continuously with time.* Digital controllers, though, involve *sampled-data variables* that are updated only at discrete sampling times. Simulations usually generate sampled-data time histories by solving *difference equations*.

As noted in Chapter 1, Section 1.1.2, difference equations, like differential equations, relate current samples of *difference equation state variables* to past samples. A dynamic system model may use

- Only difference equations (e.g., classical neural networks [1])

---

* This is true even though, as we know, the simulation time **t** actually changes in small integration steps.

- Only differential equations, like the models discussed in Chapters 1 and 2

- Both differential equations and difference equations (Chapters 3 and 4)

Difference equations complicate programming because

- A differential equation solver updates clearly identified differential equation state variables with a canned integration routine. But difference equation state variables must be explicitly selected and updated in each user program.

- Missorted defined-variable assignments are easily mistaken for extra difference equations.

- Simulations combining difference equations with differential equations must not execute difference equation code in the middle of integration steps.

In view of these problems, many simulation programs relegate difference equations to "procedural" program segments that must be written in Fortran by hopefully knowledgeable users. Desire, though, is specifically designed to deal directly with difference equations.

3.2.2 Sampled-Data Operations

Difference equation operations often (but not always; see Chapter 4, Section 4.4) execute *periodically. If the DYNAMIC program segment defining a model contains no differential equations*, the simulation run simply steps the time variable **t** through the NN uniformly spaced values

$$\mathbf{t = t0, \ t0 + COMINT, \ t0 + 2 \ COMINT,}$$
$$\mathbf{\dots , \ t0 + (NN - 1)COMINT = t0 + TMAX}$$
$$\text{with } \mathbf{COMINT = TMAX/(NN - 1)} \qquad \mathbf{(3.2)}$$

If the starting time **t0** is not specified, it defaults to 1. If the simulation-run duration **TMAX** is also left unspecified, it defaults to **NN - 1**, so that **t** simply steps through **t = 1, 2, …, NN**.

If the DYNAMIC program segment does contain differential equations, then the simulation time **t** is incremented by an integration routine, and we have the normal situation described in Chapter 1, Section 1.2.4. The initial time **t0** defaults to 0, and an error message results if **TMAX** is left unspecified. At the periodic sampling times (Equation 3.2)* Desire executes output requests and also *sampled-data assignments* following an **OUT** statement at the end of the differential equation program. *Such sampled-data assignments can implement difference equation code.*

3.2.3 Changing the Sampling Rate

The basic *sampling rate* implied by Equation 1.1 (Chapter 1) is

SR = 1/COMINT = (NN − 1)/TMAX

If it is desirable to produce time-history output at a lower rate than other sampled-data operations, we set the simulation variable **MM** to an integer value greater than 1. This causes output requests such as **dispt x, y, ...** to execute only at **t = t0** and then at every **MM**th sampling point and at **t = t0 + TMAX**. **MM** affects only the output sampling rate and leaves **NN**, **COMINT**, and **SR** unchanged.

One can also lower the sampling rate for some or all sampled-data assignments by preceding them with a **SAMPLE m** statement that follows or replaces **OUT**. Assignments following **SAMPLE m** execute only at **t = t0** and then at every **m**th sampling point and at **t = t0 + TMAX**. This technique lets you sample pseudorandom noise (Chapter 4, Section 4.3.4) at the rate **SR**, other sampled-data operations at the sampling rate **SR/m**, and output at the sampling rate **SR/MM**.

3.3 DIFFERENCE EQUATION PROGRAMMING

3.3.1 Primitive Difference Equations

The most user-friendly difference equations are *simple recurrence relations* such as

$$q = F(t; q) \qquad (3.3)$$

Desire automatically recognizes that **q** *is a difference-equation state variable and assigns it the initial value* **q(t0)**. **q(t0)** defaults to 0 unless the experiment-protocol script has assigned a different value.

As an example, the simple recurrence relation

---

* Recall from Chapter 1, Section 1.2.4, that fixed integration step sizes DT should equal integral fractions of **COMINT = TMAX/(NN − 1)**.

$$x = 2 * (x^2 - x) \quad \text{with} \quad x(1) = 0.4$$

generates pseudorandom noise. But such relatively primitive difference equations have much interesting and important applications (Chapter 4, Sections 4.4.1–4.4.5).

3.3.2 General Difference Equation Systems

More general difference equation systems relate current values of **N** difference equation state variables **q1, q2, ... , qN** to their last-computed past values by **N** difference equations

$$qi = Qi(t; q1, q2, ... ,qN; p1, p2, ...) \quad (i = 1, 2, ... , N) \quad (3.4)$$

where **p1**, **p2**, ... are *defined variables* determined, as in Chapter 1, Section 1.1.3, by properly sorted defined-variable assignments

$$pj = Pj(t; q1, q2, ... , qN; p1, p2, ...) \ (j = 1, 2, ...) \quad (3.5)$$

But difference equations in the general form (Equation 3.4) must *not* be executed as assignments! Unlike the simple recurrence relations (Equation 3.3), the equations (Equation 3.4) relate *multiple* state variables, which can be correctly updated only after *all* the expressions on the right have been computed using past values of the state variables **qi**.

A program for solving the difference equations (Equation 3.4) therefore requires *three successive sets of assignments*:

1. The *defined-variable assignments* (Equation 3.5)

2. **N** *difference equation assignments* to **N** auxiliary defined variables

$$Qi = Qi(t; q1, q2, ... ,qN; p1, p2, ...) \quad (i = 1, 2, ... , N) \quad (3.6)$$

3. **N** *updating assignments* *

$$qi = Qi \ (i = 1, 2, ... , N) \quad (3.7)$$

---

* Note that the updating assignments act much like the integration routine of a differential equation solver, with the variables **Qi** playing the part of the hidden derivatives **Gi** in Chapter 1, Section 1.2.3.

The defined-variable assignments (Equation 3.5) must be sorted into procedural order as in Chapter 1, Section 1.1.3. The difference equation assignments (Equation 3.6) must follow the defined-variable assignments but need not be ordered. Updating assignments (Equation 3.7) can follow the difference equation assignments in any order.

It is easy to overlook the fact that the state-variable values **qi** produced by the updating assignments (Equation 3.7) are *not* the **qi** values at the current sampling time **t**—they are associated with the *next* sampling time. *To display or list the time history of any difference-equation state variable, say* **q(t)**, *one must assign its value to an auxiliary defined variable* **qq = q** *before* **q** *is updated. One must then display or list* **qq(t)** *rather than* **q(t)** to ensure that state variable values as well as defined variables are output at the correct sampling times. Figure 3.6 shows a small difference equation program and a listing that illustrates these time relationships. Figure 3.7 exhibits another complete difference equation program, this time with a graphics display.

Note at this point that the experiment-control commands **reset** and **drunr** reset only **t**, **DT**, and *differential equation* state variables. *Difference equation* state variables must be reset individually.

3.3.3 Combined Systems Imply Sample/Hold Operations

3.3.3.1 Difference Equation Code and Differential Equation Code

In DYNAMIC program segments, difference equation code such as that described in Section 3.2.1 executes like any other set of assignments. *If there are no differential equations*, repeated calls to Desire's derivative routine (Chapter 1, Section 1.2.3)* now solve the difference equations instead of differential equations. The integration routine reverts to Rule 0, which merely advances the simulation time with

t = t + COMINT

Models *combining differential and difference equations* must not update difference equation variables in the middle of integration steps. For this reason, any and all difference equation code (Equation 3.5 through Equation 3.7) *must* follow an **OUT, SAMPLE m** and/or **step** statement

---

* For simplicity we still call this a derivative routine, because the program flow in Chapter 1, Figure 1.2, applies even if the DYNAMIC program segment contains only difference equations.

```
t, xx, yy, p
1.00000e+000  5.00000e+000  1.00000e+001  5.00000e+000
2.00000e+000  8.00000e+000  1.10000e+001  8.00000e+000
3.00000e+000  1.10000e+001  1.20000e+001  1.10000e+001
4.00000e+000  1.40000e+001  1.30000e+001  1.40000e+001
5.00000e+000  1.70000e+001  1.40000e+001  1.70000e+001
6.00000e+000  2.00000e+001  1.50000e+001  2.00000e+001
7.00000e+000  2.30000e+001  1.60000e+001  2.30000e+001
8.00000e+000  2.60000e+001  1.70000e+001  2.60000e+001
9.00000e+000  2.90000e+001  1.80000e+001  2.90000e+001
1.00000e+001  3.20000e+001  1.90000e+001  3.20000e+001
```

```
--              SIMPLE DIFFERENCE EQUATIONS
-----------------------------------------------------------------
NN = 10
X = 5 | y = 10 | --              initial values at t = 1
Drun
-----------------------------------------------------------------
DYNAMIC
-----------------------------------------------------------------
p = x | --                        a defined variable
xx = x | yy = y | --      defined variables for output
X = x + 3 | --                 2 difference equations
Y = y + 1
X = X | y = Y | --            updating assignments
-----------------------------------------------------------------
type xx, yy, p
```

FIGURE 3.6   A simple difference equation program and the resulting listing. The auxiliary defined variables **xx** and **yy** correctly produce values of the state variables **x** and **y** before updating.

placed at the end of the differential equation code. All code following **OUT** or **SAMPLE m** executes periodically at sampling points and models *periodic sampled-data operations*. **step** postpones execution of all succeeding code until the end of the current integration step; suitable programs will be discussed in Chapter 4, Sections 4.3.5 and 4.4.1 through 4.4.4.

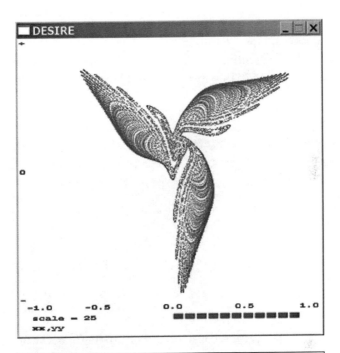

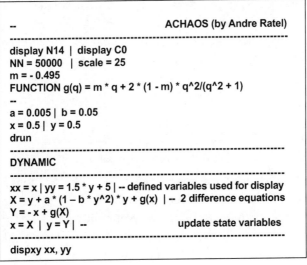

ACHAOS (by Andre Ratel)

```
--                                    ACHAOS (by Andre Ratel)
-----------------------------------------------------------------------------
display N14 | display C0
NN = 50000  | scale = 25
m = - 0.495
FUNCTION g(q) = m * q + 2 * (1 - m) * q^2/(q^2 + 1)
--
a = 0.005 | b = 0.05
x = 0.5 | y = 0.5
drun
-----------------------------------------------------------------------------
DYNAMIC
-----------------------------------------------------------------------------
xx = x | yy = 1.5 * y + 5 | -- defined variables used for display
X = y + a * (1 – b * y^2) * y + g(x) | -- 2 difference equations
Y = - x + g(X)
x = X | y = Y | --                            update state variables
-----------------------------------------------------------------------------
dispxy xx, yy
```

FIGURE 3.7   The difference-equation program in this figure created **NN** = 50,000 display points (program contributed by Andre Ratel).

3.3.3.2 Transferring Sampled Data

Significantly,

- Difference equation assignments can read variables computed by the preceding differential equation code as difference equation defined variables. Such variables typically represent the output of analog-to-digital converters.

- Feeding *difference equation state variables* back to preceding differential equation assignments does not cause undefined-variable errors, because state variables necessarily have initial values (they are either set by the experiment-protocol script or default to 0).

- *Difference equation defined variables* fed back to the differential equation system are *sample/hold state variables (track/hold state variables)* and must be assigned initial values by the experiment protocol. Desire returns an "undefined variable" error message if you forget this.

3.3.3.3 Simulation of Sampled-Data Reconstruction

One can readily simulate reconstruction of "analog" data from sampled data by feeding it to a low-pass filter (e.g., to simulate "analog" noise, Chapter 4, Section 4.3.4), or by interpolating or extrapolating samples from two or more delayed sampled-data sequences. But it is impossible to display or list interpolated data when display and sampled-data sampling rates are equal. To demonstrate interpolation, one must slow the sampled-data sampling rate with **SAMPLE m** (Chapter 3, Section 3.2.3).

3.4 A SAMPLED-DATA CONTROL SYSTEM

3.4.1 Simulation of an Analog Plant with a Digital PID Controller

The programming techniques described in Chapter 3, Sections 3.2 and 3.3, were specifically developed to permit simulation of sampled-data control systems. Figure 3.8 models an *analog plant* with a differential equation system and the periodic sampled-data operations of a *digital controller* with difference equations. Difference equation assignments follow an **OUT** and/or **SAMPLE m** statement at the end of the differential equation code. Sampled-data operations then execute at the uniformly spaced sampling times (Equation 3.2).

Analog-plant variables fed to sampled-data assignments model *analog-to-digital converters with track/hold inputs*. Difference equation variables

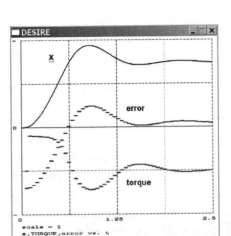

```
-- SERVO WITH DIGITAL PID CONTROLLER
------------------------------------------------------------
TMAX = 2.5 | DT = 0.001 | NN = 700
display N14 | display C7 | display R
--
TS = 0.05 | -- simulated sampling rate is 1/TS
m = TS * (NN - 1)/TMAX
------------------------------------------------------------
u = 0.7 | -- step input
maxtrq = 0.8 | R = 3 | -- motor parameters
------------------------------------------------------------
--                 initial t, x, xdot default to 0
x1 = 0 | x2 = 0 | -- initialize digital controller ...
y = 0 | --              ... and y, too!
--
KP = 3 | KI = 1.2 | KD = 0 .2 | -- PID parameters
B0 = KD/TS | B1 = - KP + 0.5 * KI * TS - 2 * KD/TS
B2 = KP + 0.5 * KI*TS + KD/TS
drun
------------------------------------------------------------
DYNAMIC
------------------------------------------------------------
torque = maxtrq * tanh(y/maxtrq) | -- analog plant
d/dt x = xdot | d/dt xdot = 10 * torque - R * xdot
--
SAMPLE m | --                 digital controller
error = x - u
y = B0*x1+ B1*x2 + B2*(x2-error) | -- controller output
X1 = x2 | X2 = x2 - error | - difference equations
--
x1 = X1 | x2 = X2 | --    update state variables
------------------------------------------------------------
TORQUE = 0.5 * (torque - scale) | -- offset display
dispt x, TORQUE, error
```

FIGURE 3.8 Simulation of a sampled-data control system. The analog plant variables are updated at appropriately small integration steps, but the digital-controller variables following the **SAMPLE m** statement are updated only at the specified sampling points, just as in a real digital control system.

fed back to the differential equation system represent *digital-to-analog converter outputs*.

The program in Figure 3.7, simulates a *digital PID (proportional/integral/derivative) controller* [2,3] that governs a motor and load modeled with differential equations exactly as in Section 3.1.2,

torque = maxtrq * tanh(y/maxtrq) | -- analog plant
d/dt x = xdot | d/dt xdot = 10 * torque - R * xdot

The torque-controlling signal **y** is now produced by the simulated digital controller. **y** is a sample/hold state variable (Section 3.3.3) and must be given an initial value by the experiment-protocol script. The program neglects analog-to-digital-converter quantization, but this could be easily modeled in the manner of Figure 4.7 (Chapter 4).

References 2 and 3 present comprehensive introductory discussions of PID control. Our digital PID controller z-transfer function is

$$G(z) = KP + \tfrac{1}{2}(KI + TS) \frac{z+1}{z-1} + \frac{KD(z-1)}{TS\,z} + \frac{Az2 + Bz + C}{z(z-1)}$$

where **KP, KI**, and **KD** are the proportional, derivative, and integral gain parameters [2,3]. Referring to Figure 3.8, the experiment-protocol script precomputes the coefficients

B0 = KD/TS B1 = - KP + 0.5 * KI * TS - 2 * B0
B2 = KP + 0.5 * KI * TS + B0

In the DYNAMIC program segment of Figure 3.8, **SAMPLE m** causes all sampled-data variables to be updated at every **m**th communication point; we set **m = TS * (NN-1)/TMAX**, where **1/TS** is the *sampling rate*. The simulated digital controller (lines following **SAMPLE m** in Figure 3.8) reads the control system output **x** with an analog-to-digital converter and computes the sampled-data error **error = x − u**, where **u** is the desired output. The program then produces the controller output **y** by solving the difference equation system

error = x - u
y = B0 * x1+ B1 * x2 + B2 * (x2 - error) | -- controller output

X1 = x2 | X2 = x2 – error | -- difference equations

--

x1 = X1 | x2 = X2 | -- update state variables

exactly as a real digital controller would.

REFERENCES

1. Korn, G.A. (2007), *Advanced Dynamic-System Simulation: Model Replication Techniques and Monte Carlo Simulation*, Wiley, Hoboken, NJ.
2. Dorf, R.C., and R.H. Bishop (2008), *Modern Control Systems*, 11th ed., Pearson Prentice-Hall, Upper Saddle River, NJ.
3. Franklin, G.F. et al. (1990), *Digital Control of Dynamic Systems*, Addison-Wesley, Reading, MA.
4. Ogata, K. (2010), *Modern Control Engineering*, 5th ed., Pearson Prentice-Hall, Upper Saddle River, NJ.

CHAPTER 4

Function Generators and Submodels

4.1 OVERVIEW

4.1.1 Introduction

This chapter explains the different function assignments used in our simulation programs, and introduces the definition and invocation of submodels.

Besides standard library functions such as trigonometric functions, users can construct functions by interpolation in stored tables (Section 4.2.2) and define their own new functions as mathematical expressions (Section 4.2.3). Section 4.3 deals with limiters, switches, and noise generators and discusses the serious problem of integrating such functions. Section 4.4 describes simple recurrence relations (difference equations) that model track/hold circuits, signal generators, backlash and hysteresis, and inverse functions. Sections 4.5 and 4.6 introduce a very efficient implementation of submodels and discuss switching in control-system simulations.

4.2 GENERAL-PURPOSE FUNCTION GENERATION

4.2.1 Library Functions

Both experiment-protocol scripts and DYNAMIC program segments admit the *library functions*

sqrt(x) sin(x) cos(x) tan(x) exp(x)

asin(x) acos(x) atan(x) sinh(x) cosh(x) tanh(x)
atan2(y,x) ≡ arctan(y/x) between $-\pi$ and $+\pi$
sinc(x) ≡ sin(x)/x
ln(x) (base *e*) **log(x)** (base 10)
sigmoid(x) ≡ 1/(1 + e $^{-x}$)
SIGMOID(x) ≡ 0, x²/(1 + x²) respectively for **x ≤ 0, x > 0**
ran()   gauss(0) (64-bit Linux only)
abs(x) ≡ |x| lim(x) sat(x) SAT(x) deadz(x)
tri(x) ≡ 1 - |x| round(x)
swtch(x) sgn(x) deadc(x) rect(x)
comp(x, minus, plus)

sigmoid(x) and **SIGMOID(x)** are used mainly in neural network models [1]. The piecewise-linear functions in the last three lines are discussed in Sections 4.3.1 through 4.3.3. Function arguments can be user-written expressions.

Additionally, experiment-protocol scripts (but not DYNAMIC segments) admit the real functions

trnc(x) (truncation to the nearest integer with a lower absolute value)
tim(0) (returns the time in seconds)

trnc(x) and **round(x)** produce integer-valued real numbers, *not* integers.

4.2.2 Function Generators Using Function Tables

4.2.2.1 Functions of One Variable

In DYNAMIC program segments, the assignment

$$Y = func1(X; F) \tag{4.1}$$

produces a continuous function **Y = y(X)** of **X** by linear interpolation* in a table that specifies **n** function values **y[1]**, **y[2]**, ..., **y[n]** corresponding to **n** argument values **x[1]**, **x[2]**, ..., **x[n]** (breakpoint abscissas). To define the function table, the experiment-protocol script declares an array **F** that concatenates an array **x** of the **n** breakpoint abscissas **x[i]** and an array **y** of the **n** function values **y[i]**, as shown in Chapter 5, Section 5.3.1.2.

---

* Desire actually implements the interpolation with fuzzy logic [3].

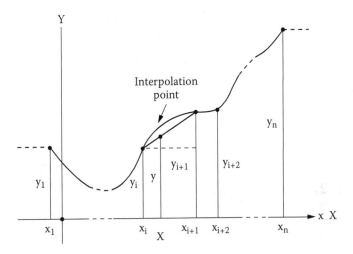

FIGURE 4.1 The DYNAMIC-segment statement Y = func1(X; F) approximates a function Y = y(X) of the argument X by table lookup and linear interpolation in a table of breakpoint coordinates xi = x[i] and yi = y[i] (i = 1, 2, . . . , n). The function generator implements the relation in Equation 4.2.

The function-table entries can be read from a **data** list or file (Chapter 5, Section 5.3.2.2), or the experiment-protocol script can compute them. Figure 4.1 illustrates the interpolation scheme. The assignment (Equation 4.1) produces the function **Y = y(X)** defined by

$$Y = y[i] + \frac{y[i+1] - y[i]}{x[i+1] - x[i]} (X - x[i])$$

$$(i = 1, 2, \dots, n - 1) \text{ for } x[1] \le X \le x[n \quad (4.2a)$$

Beyond the range of the table we use

$$X \le x[1] \text{ for } X \le x[1] \qquad Y = y[n] \text{ for } X \ge x[n] \qquad (4.2b)$$

Figure 4.2 shows an example.

4.2.2.2 Functions of Two Variables

The DYNAMIC-segment assignment

$$f = func2(X, Y; F, x, y)$$

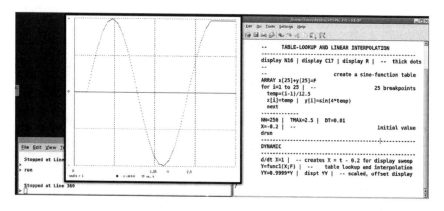

FIGURE 4.2   In this program, the experiment protocol employs a program loop (Sec. 5.2.3) to create a sine-function table. The DYNAMIC program segment generates an display-abscissa sweep **X = t - 0.2** and uses the table-lookup/interpolation function generator **func1(X; F)** to produce the function plot.

computes the function **f** by bilinear interpolation* in a two-dimensional function table of nm values

$$F[i, k] = f(x[i], y[k]) \quad (i = 1, 2, \dots , n; k= 1, 2, \dots , m)$$

The experiment protocol must declare and fill an array **x** of **n** X-breakpoints **x[i]**, an array **y** of **n** Y-breakpoints **y[k]**, and an array **F** of **nm** function values **F[i, k]**,

   ARRAY x[n], y[m], F[n, m], ...

func2 returns an error if the array dimensions do not match.

4.2.2.3 General Remarks

A function table can be used for multiple invocations of the same function with different arguments, for example,

   Y1 = func1(X1; F, x, y)   Y2 = func1(X2; F, x, y)

See also the user-program examples **func1-black.src, func2.src** on the book CD.

---

* Desire again implements the interpolation with fuzzy logic.

TABLE 4.1   Examples of User-Defined Functions

FUNCTION cotan(x$) = cos(x$)/sin(x$)
FUNCTION vabs(u$, v$) = sqrt(u$^2 + v$^2)
FUNCTION min(a$, b$) = a$ - lim(a$ - b$)
FUNCTION max(aa, b$) = a$ + lim(b$ - a$)
FUNCTION fswitch(x$, c$, m$, p$) = m$ + (p$ - m$) * swtch(x$ - c$)

Note: Such functions can be collected in library files for reuse.

4.2.3 User-Defined Functions

Experiment-protocol scripts create user-defined functions with **FUNCTION** declarations such as

FUNCTION abs2 (u$, $v) = sqrt(u$^2 + v$^2)

Both experiment-protocol scripts and DYNAMIC program segments can invoke user-defined functions with new arguments, so that

z = abs2(x,y)   is equivalent to   **z = sqrt(x^2 + y^2)**

This produces fast in-line code *without any function call/return overhead.* Invocation arguments can be expressions. Table 4.1 shows some examples.

A function definition must fit one program line—but this can be a long line "wrapping around" on the display. Function definitions may include constant parameters and variables. We marked the *dummy arguments* **x$**, **y$**, **z$** with dollar signs, but that is not necessary. Dummy arguments must not be subscripted. Dummy argument names are "protected"; that is, any attempt to use their names after the function definition produces an error message.

FUNCTION definitions can be used to give conveniently readable names to table-lookup functions, for example,

FUNCTION spider(x$, y$) = func2(x$, y$; F, X, Y)

FUNCTION definitions can be *nested;* that is, they can contain previously defined functions. But *recursive definitions* such as

FUNCTION f1(x) = f1(x) + 1

or

FUNCTION f1(x) = f2(1) + x | FUNCTION f2(y) = f1(y+1)

and also recursive function calls, as in

FUNCTION incr(x) = x + 1 | q = incr(incr(incr(y)))

are illegal.

4.3 LIMITERS AND NONCONTINUOUS FUNCTIONS

4.3.1 Limiters

4.3.1.1 Introduction
The piecewise-linear library functions in Figure 4.3 are useful for constructing device transfer characteristics and special switches in engineering applications. These functions work both in experiment-protocol programs and in DYNAMIC program segments.

4.3.1.2 Simple Limiters
lim(x) is a *simple limiter* (half-wave rectifier), and **sat(x)** is a unit-gain *saturation limiter* that limits its output between −1 and 1. **SAT(x)** similarly limits its output between 0 and 1. More general saturation limiters are modeled with

y = a * sat(x/a)   (gain 1, limits between - a and a > 0)
y = lim(x - min) - lim(x - max)   (gain 1, limits between min and max > min)

Many continuous functions can be approximated as sums of simple limiter functions,

a0 + a1 * lim(x - x1) + a2 * lim(x - x2) + . . .

The library functions **deadz(x)** (*deadspace function*, Figure 4.3) and **tri(x)** (*triangle function*) are defined by

deadz(x) ≡ x - sat(x) ≡ x - ½ [abs(x + 1) - abs(x - 1)]
tri(x) ≡ 1 - abs(x)

The modified triangle function **lim[tri(x)]** in Figure 4.3 is also often used.

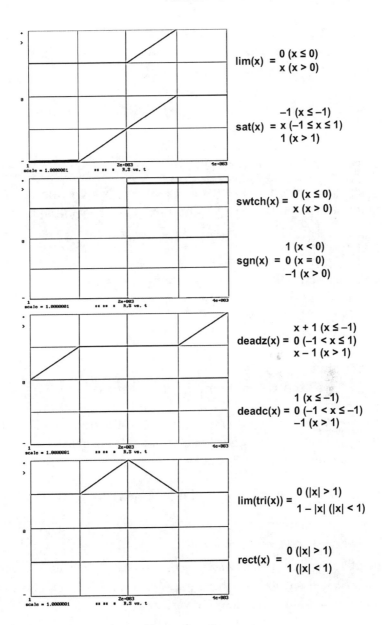

$$\text{lim}(x) = \begin{array}{l} 0 \ (x \le 0) \\ x \ (x > 0) \end{array}$$

$$\text{sat}(x) = \begin{array}{l} -1 \ (x \le -1) \\ x \ (-1 \le x \le 1) \\ 1 \ (x > 1) \end{array}$$

$$\text{swtch}(x) = \begin{array}{l} 0 \ (x \le 0) \\ x \ (x > 0) \end{array}$$

$$\text{sgn}(x) = \begin{array}{l} 1 \ (x < 0) \\ 0 \ (x = 0) \\ -1 \ (x > 0) \end{array}$$

$$\text{deadz}(x) = \begin{array}{l} x + 1 \ (x \le -1) \\ 0 \ (-1 < x \le 1) \\ x - 1 \ (x > 1) \end{array}$$

$$\text{deadc}(x) = \begin{array}{l} 1 \ (x \le -1) \\ 0 \ (-1 < x \le -1) \\ -1 \ (x > 1) \end{array}$$

$$\text{lim(tri}(x)) = \begin{array}{l} 0 \ (|x| > 1) \\ 1 - |x| \ (|x| < 1) \end{array}$$

$$\text{rect}(x) = \begin{array}{l} 0 \ (|x| > 1) \\ 1 \ (|x| < 1) \end{array}$$

FIGURE 4.3   Piecewise-linear library functions.

4.3.1.3 Useful Relations between Limiter Functions

In most digital computers, the fastest nonlinear operation is not the simple limiter (half-wave rectifier) but the absolute-value function (full-wave rectifier) **abs(x)** ≡ **|x|,** which simply resets the sign bit. It is, therefore useful to remember the identities

$$\mathrm{lim}(x) \equiv \tfrac{1}{2}\,(x + |x|) \equiv \tfrac{1}{2}\,x + |\tfrac{1}{2}\,x|$$
$$\mathrm{sat}(x) \equiv \mathrm{lim}(x + 1) - \mathrm{lim}(x - 1) \equiv \tfrac{1}{2}\,[\mathrm{abs}(x + 1) - \mathrm{abs}(x - 1)]$$
$$\mathrm{SAT}(x) \equiv \mathrm{lim}(x) - \mathrm{lim}(x - 1) \equiv \tfrac{1}{2}\,[1 + \mathrm{abs}(x) - \mathrm{abs}(x - 1)]$$

It is often possible to get rid of the numerical factor ½ by rescaling.*

4.3.1.4 Maximum and Minimum Functions
The largest and smallest of two arguments x, y are

$$\mathrm{max}(x, y) \equiv x + \mathrm{lim}(y - x) \equiv y + \mathrm{lim}(x - y) \equiv \tfrac{1}{2}\,[x + y + \mathrm{abs}(x - y)]$$
$$\mathrm{min}(x, y) \equiv x - \mathrm{lim}(x - y) \equiv y - \mathrm{lim}(y - x) \equiv \tfrac{1}{2}\,[x + y - \mathrm{abs}(x - y)]$$

Note also

$$\mathrm{lim}(x) \equiv \mathrm{max}(x, 0) \qquad \mathrm{max}(x, y) - \mathrm{min}(x, y) \equiv x + y$$

4.3.1.5 Output-Limited Integrators
The differential equation

$$dy/dt = \mathrm{swtch}(\mathrm{max} - y)\,\mathrm{lim}(x)$$
$$+\ \mathrm{swtch}(y - \mathrm{min})\,\mathrm{lim}(-x) \quad (\mathrm{min} < \mathrm{max})$$

models an *output-limited integrator* that stops integrating when the output **y** exceeds preset bounds. Note that this is quite different from an *integrator followed by an output limiter*,

$$dY/dt = x \mid y = \mathrm{lim}(Y - \mathrm{min}) - \mathrm{lim}(Y - \mathrm{max}) \quad (\mathrm{min} < \mathrm{max})$$

4.3.2 Switches and Comparators
Referring to Figure 4.3, the *switch function* **swtch(x)** switches from 0 to 1 for **x > 0**, and **sign(x)** switches from –1 to 1. Switch functions can be combined in many different ways. In particular,

$$u = \mathrm{swtch}(t - t1) - \mathrm{swtch}(t - t2) \quad (t1 < t2)$$

produces a *unit pulse* **u = u(t)** starting at **t = t1** and ending at **t = t2**. To turn a system input **v(t)** on at **t = t1** and then turn it off at **t = t2**, simply replace **v** by **v * u**.

---

* In fact, Desire's library functions are implemented in this way.

swtch(x) and **sgn(x)** are *comparators* whose output changes when the input **x** crosses zero. **deadc(x)** is a *comparator with deadspace* between **x = - 1** and **x = 1**. The library function

$$\textbf{comp(x, minus, plus) =} \quad \begin{matrix} \textbf{plus (x > 0)} \\[6pt] \textbf{minus (x < 0)} \end{matrix}$$

models a *relay comparator* or *function switch*: **y** switches between the values **minus** and **plus** when the input variable **x** crosses 0. A relay comparator can also be implemented with

y = minus + (plus - minus) swtch(x)

Similarly, the function

y = minus * swtch(a - x - delta) + plus * swtch(x - a - delta)

represents a relay comparator with the symmetrical deadspace **±delta**.

Note that time-history displays cannot correctly display switched functions that switch more than once between display sampling points; you may want to increase the number of display points (**NN** or **NN/MM**; Chapter 3, Section 3.2.3) as needed.

4.3.3 Signal Quantization

Simulations of digital controllers, signal-processing systems, and instrumentation often require quantization of continuous signals **x(t)**. The short program in Figure 4.4 employs the library function **round(x)** to quantize the sinusoidal output **x(t)** of a harmonic oscillator modeled with a pair of state equations. **y(t)** is the quantizer output, and **a** is the *quantization interval*. The quantizer error **y - x** is often called *quantization noise*. Note that **round(x)** produces a quantized floating-point variable, not an integer.

4.3.4 Noise Generators

The library function **ran()** produces a sample of pseudorandom noise uniformly distributed between −1 and 1. **ran()** works in both experiment-protocol scripts and DYNAMIC program segments. Under Linux only, **gauss(0)** generates a sample of pseudorandom Gaussian noise with mean

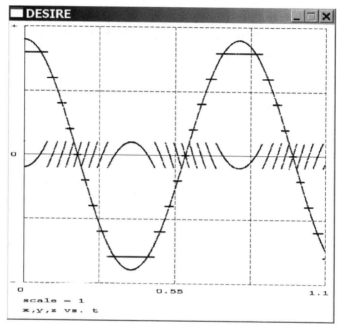

TMAX = 1.1 | DT = 0.001 | NN = 1001
w = 8 | x = 0.9 | --   quantization interval
a = 0.2
drun

DYNAMIC
----
d/dt x = w * xdot | d/dt xdot = – w * x | --   signal
step
y = a * round(x/a) | --   quantized signal
z = y – x | --   quantizer error (quantization noise)
dispt x, y, z

FIGURE 4.4   Demonstration of signal quantization and quantization noise.

0 and variance 1. To obtain such Gaussian samples under Windows, use the Box–Muller-type formula

X = sqrt(- 2 * ln(abs(ran()))) * cos(2* PI * abs(ran()))

Simulation programs assume that noise generators produce statistically independent samples. But that is really not true. Pseudorandom-noise samples, although usually guaranteed to be uncorrelated, are generated by a deterministic program. Reference 1 discusses tests of pseudorandom-noise quality. But we usually *assume* statistical independence and then compare results obtained with different pseudorandom-noise algorithms.

Desire simulation of systems with random inputs and/or random parameters, and computation of the resulting random-process statistics, are described in a separate text [1]; our book CD has examples. Readers of this introductory text, though, ought to be aware of the following points:

- **ran()** and **gauss(0)** are most frequently used for *periodic sampled-data operations* preceded by **OUT** or **SAMPLE m** (Chapter 3, Sections 3.2.2 and 3.2.3). This results in predictable noise power spectra [1] and also prevents execution of these discontinuous functions in the middle of integration steps (Section 4.3.5).*

- Sampled periodic noise, say **x = ran()**, feeding a differential equation system (say, a simulated analog filter or control system) is necessarily a track/hold state variable and must be initialized by the experiment protocol (Chapter, Section 3.3.3).

4.3.5 Integration through Discontinuities and the Step Operator

Switches, comparators, noise generators, and rounding are discontinuous functions, and limiter functions have noncontinuous derivatives. *Such operations must not be allowed to execute in the middle of integration steps, for all integration rules other than the Euler rule* (Chapter 1, Section 1.2.5.2) *require "smooth" integrands with continuous derivatives.*

It is, in principle, possible to replace limiters and switching functions with differentiable approximations. **sat(x)**, for instance, can be approximated by **tanh(a * x)** with a large value of the parameter **a**. But this introduces large absolute Jacobian eigenvalues (stiffness, see Section A.2.3 in the Appendix) and thus requires more complicated integration rules and very small integration steps.

Early simulation projects ignored the problem altogether and often produced reasonable results, either by sheer luck or perhaps because stable control system models reduced errors. Improved simulation languages *predict* the nearest switching time **t = Tevent** of, say, **swtch(x)** by using past and present values of the argument **x** to extrapolate future values. The integration routine must then force the nearest integration step to end at **t = Tevent**. The extrapolation formula has to be as accurate as the current integration

---

* Nonperiodic noise would require a **step** statement (Section 4.9).

rule, and the program must select the first function likely to switch [2]. The resulting computations can be cumbersome, especially if there are several switches and/or limiters.

Desire uses a simplified approach with its **OUT, SAMPLE m**, and **step** statements. In simulations combining differential equations and periodic sampled-data operations, limiters, switches, noise, and rounding functions *called by sampled-data operations* are safe, because they necessarily follow an **OUT** or **SAMPLE m** statement (Chapter 1, Section 1.2.4.2 and Chapter 3, Section 3.3.3.1). But if such functions are *called within a differential equation system*, they must follow a **step** statement that postpones their execution until the end of the current integration step. Switching-time accuracy then requires acceptably small integration steps near a switching point.

With variable-step integration rules, the maximum step size **DTMAX** must be set to a small value. With the fixed-step integration rules 1, 2, and 3 (Table A.1 in the appendix), one sets **DT** to a small integral fraction of **COMINT = TMAX/(NN − 1)** (Chapter 1, Section 1.2.4.1). Integration rules 2, 3, and 5 let the users themselves program the integration step size during simulation runs by assigning new **DT** values in a DYNAMIC program segment. When you use, say, **swtch (x − a)**, you can decrease **DT** as a function of **|x - a|** (Section 4.6.2).

4.4 VERY USEFUL MODELS EMPLOY SIMPLE RECURRENCE RELATIONS

4.4.1 Introduction

Section 3.3.1 (Chapter 3) discussed models using *simple difference equations*

$$q = F(t; q)$$

for periodic sampled-data operations, as in digital control systems. We shall now apply such recursive assignments within "analog" differential equation systems to model a number of widely useful operations.

As noted in Chapter 3, Section 3.3.1, Desire automatically recognizes **q** as a difference equation state variable and assigns it the default initial value 0. As we saw in Chapter 3, difference equation state variables are not automatically reset by **reset** or **drunr** statements but must be reset explicitly by the experiment protocol.

If the function **F(t; q)** involves limiters or switches (as in the following examples), then the difference equation must follow a **step**, **OUT**, or **SAMPLE m** statement to ensure correct integration (Section 4.3.5).

4.4.2 Track/Hold Circuits and Maximum/Minimum Tracking

Figure 4.5a indicates how the recursive assignment

y = y + (x - y) * swtch(ctrl)

models a *track/hold circuit*. **y** *tracks* **x** when **ctrl > 0** and *holds* its last value when **ctrl** ≤ 0.

The recursive assignment,

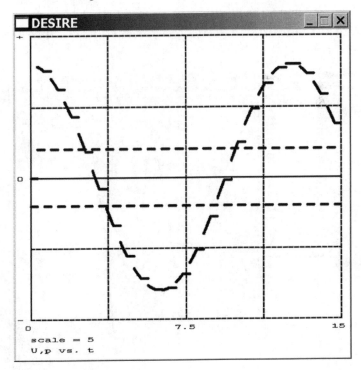

FIGURE 4.5a   Track/hold operation obtained with the recurrence relation
y = y + swtch(ctrl) * (x - y).
y tracks the sine wave **x** when **ctrl > 0** and holds its last value when **ctrl ≤ 0**. The control pulses were generated with the program of Figure 4.9, but a sine function would do as well.

max = x + lim(max - x)

tracks and holds the largest past value of a time function **x(t)** if the experiment protocol initializes the difference equation state value **max** with a large negative value such as **1.0E+30**. Similarly,

min = x + lim(x - min)

tracks and holds the smallest past value of **x(t)** if the experiment protocol initializes **min** with a large positive value such as **1.0E+30**. Again,

MAX = abs(x) + lim(MAX - abs(x))

tracks and holds the largest absolute value of **x(t)**, with **MAX** initialized to **- 1.0E+30**. Figure 4.5b illustrates these operations, which are useful for modeling instrumentation and also for automatic display scaling (Chapter 6, Section 6.8.1).

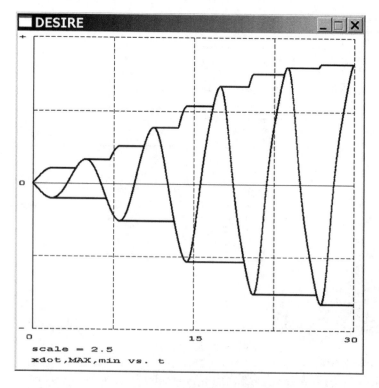

FIGURE 4.5b   Maximum-absolute-value and minimum-value tracking.

4.4.3 Models with Hysteresis

4.4.3.1 Simple Backlash and Hysteresis

Figure 4.6 models simple *gear backlash* with the recursive assignment

y = y + a * deadz((x - y)/a)

Suitably chosen functions of **y**, such as **tanh(10 * y)** can be similarly used to model simple hysteresis (Figure 4.7). More realistic hysteresis models must be derived directly from physics and usually involve differential equation systems as well as difference equations.

4.4.3.2 A Comparator with Hysteresis

The recursive assignment

p = A * sgn(p - x)

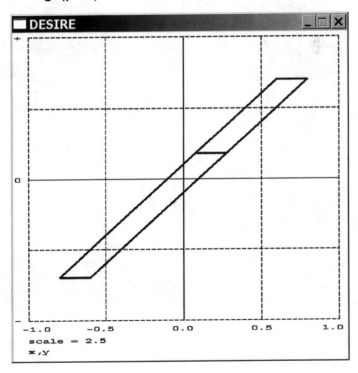

FIGURE 4.6   Simulation of simple backlash. As with all hysteresis models, this is not a static transfer characteristic but implements a difference equation.

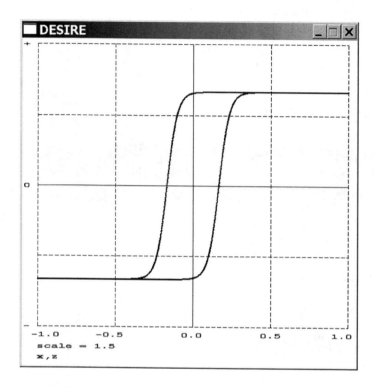

FIGURE 4.7 This transfer characteristic was derived from that of Figure 4.6 by adding the defined variable **z = tanh(c * y)**.

is the digital equivalent of an analog comparator with regenerative feedback (*Schmitt trigger*, Figure 4.8). **p** is a difference equation state variable, whose initial value is normally set equal to the parameter **A**.

4.4.3.3 A Deadspace Comparator with Hysteresis

As another example, the recursive assignment

y = deadc(A * y - x)

models a *deadspace-comparator with hysteresis* (Figure 4.9). This is useful for modeling the operation of paired space-vehicle attitude-control rockets (Section 4.6.1).

4.4.4 Signal Generators

4.4.4.1 Square Wave, Triangle, and Sawtooth Waveforms

Integrator feedback around a Schmitt trigger produces a widely useful *signal generator for square waves and triangle waves*, just as it does in the

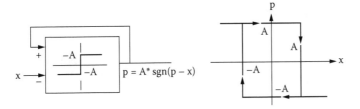

FIGURE 4.8   The recursive assignment **p = sgn(p - x)** acts exactly like an analog comparator with regenerative feedback (Schmitt trigger).

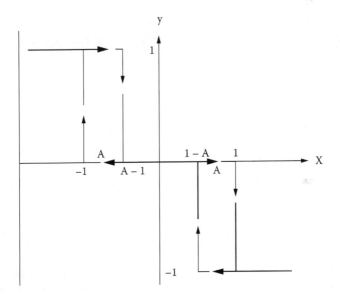

FIGURE 4.9   A deadspace/hysteresis transfer characteristic similar to that associated with many vernier attitude-control rockets.

analog world (Figure 4.10). Figures 4.11a and 4.11b show a signal generator program and the resulting waveforms. The program lines

d/dt x = a * p
step
p = A * sgn(p - x)

generate periodic square waves **p = p(t)** and triangle waves **x = x(t)**, both of amplitude **A** and period **4A/a**. We give **x** the initial value **A**, and let the initial value of the difference equation state variable **p** default to 0.

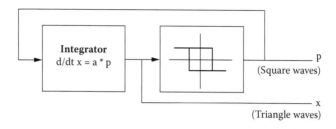

FIGURE 4.10   Integrator feedback around a comparator with hysteresis (Schmitt trigger) produces square waves **p** and triangle waves **x**.

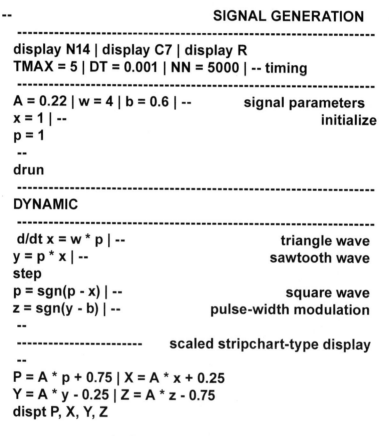

FIGURE 4.11a   Program for digital generation of useful periodic signals.

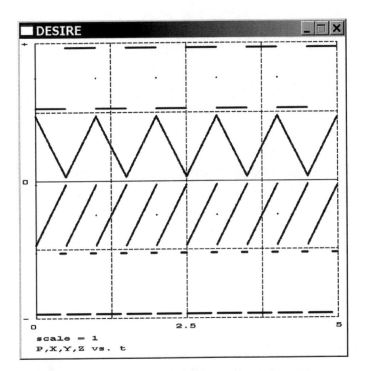

FIGURE 4.11b   Waveforms produced by the program of Figure 4.11a.

Next,

$$y = p * x$$

produces a *sawtooth waveform* **y** with amplitude **A** and frequency **a/2A**. Triangle and sawtooth waveforms are useful *test inputs* for trying various function generators. You can combine the sawtooth wave **y** and a non-linear-function generator—say a table-lookup generator (Section 4.2.2) to produce a wide variety of periodic waveforms.

4.4.4.2 Signal Modulation

One can *frequency-modulate* the square wave, triangle, and sawtooth waveforms of Section 4.4.4.1 by making the parameter **a** variable. You can also add a variable bias **- b** to the sawtooth **y**, and send the result to a comparator to produce *width-modulated pulses*

$$z = sgn(y - b)$$

as shown in Figure 4.11b.

Sinusoidal signals such as

s = A * sin(w * t + phi)

can be *amplitude-, frequency-, and/or phase-modulated* by variable parameters **a**, **w**, and/or **phi**.

4.4.5 Generation of Inverse Functions

DYNAMIC program segments can often compute the ***inverse* x = g(y)** *of a function* **y = f(x)** by steepest descent minimization of the squared error **E = |y - f(x)|²**. To this end, we define an error function

error(t) = y - f(x)

and invoke the simple recurrence relation

y = y - gain · error ($\partial E/\partial x$)

x and **y** can be sampled-data variables or "continuous" variables in a differential equation system. In the latter case, this recurrence relation must be invoked *even in the middle of integration steps.* The optimization gain **gain** is a positive constant adjusted as a compromise between accuracy and numerical stability.

The recursion relation tries to move **y** in the direction of the negative gradient of the squared error **E**. With suitably differentiable functions, the optimization succeeds if **y** is never far from the correct minimum. To ensure this, it is best to precompute the correct initial value of the difference equation state variable **y = g(x)** for **x = x(t0)**.

As an example,

error = y * y * y * y - x
y = y - 0.5 * y * y * y * error

causes **y** to approximate the 4th root of a positive variable **x** (example programs **rootx.src** and **root4.src**, which also compare **y** and the function **x$^{0.25}$**).

Errors in such approximations depend on the signal amplitude and frequency and need to be carefully checked in each case. Errors increase with the integration step **DT** in the case of "continuous" variables or with the sampling interval **COMINT** in the case of sampled-data variables.

4.5 SUBMODELS CLARIFY SYSTEM DESIGN

4.5.1 Submodel Declaration and Invocation

4.5.1.1 Submodels

Like user-defined functions, *submodels* in DYNAMIC program segments provide programming language extensions. Submodels are *declared* in the experiment-control script and *invoked* in a DYNAMIC program segment. Submodels can invoke multiple instances of complete differential equation systems or data-processing operations, with the same or different invocation arguments. This is much more than a convenient shorthand notation. Submodels can make simulations easier to understand, not just easier to program. Useful submodels can be collected in library files for reuse. Desire submodel invocations compile into inline code and do not cause any function call/return overhead.

Submodel definitions may invoke other submodels (*nested* submodels). But nested and recursive submodel definitions, and also recursive submodel invocations, are illegal (see also Section 4.2.3).

4.5.1.2 Submodel Declaration

Before a submodel is used in a DYNAMIC program segment, it must be defined with a **SUBMODEL** declaration in the experiment-protocol script. For example,

```
SUBMODEL quad(x$, y$, ydot$, a$, b$)
    d/dt y$ = ydot$
    d/dt ydot$ = x$ - a$ * y$ - b * ydot$
    end
```

defines a differential equation submodel representing a mass restrained by a spring and viscous friction. Submodel definitions can contain any legal DYNAMIC-segment code, including calls to user-defined functions and "global" variables common to all invocations. Program displays and listings automatically indent (prettyprint) the definition lines.

In our example, **x$, y$, ydot$, a$, b$** are *dummy arguments*, much like those in **FUNCTION** declarations (Section 4.2.3). We terminated the dummy arguments with **$** for clarity, but this is not necessary. Once a dummy argument name is used in a submodel declaration, it can no longer be used elsewhere; it is also "protected" by an error message to prevent side effects.

4.5.1.3 Submodel Invocation and Invoked State Variables

Once declared, the submodel can be *invoked* in a DYNAMIC program segment, say, with

invoke quad(input, y, ydot, ww, r)

with appropriate *invocation variables* (in this case **input, y, ydot, r, ww)** substituted for each dummy argument. Invocation arguments are called by name and so cannot be literals, expressions, or subscripted variables. The program returns an error when declaration and invocation argument lists do not match.

In our example, the submodel invocation generates compiled inline differential equation code for

d/dt y = ydot
d/dt ydot = input - ww * y - r * ydot

Note that the submodel invocation has created new state variables **y** and **ydot**. Such invocation state variables need to be declared in the experiment protocol script with a **STATE** declaration, in our case

STATE y, ydot

4.5.2 A Simple Example: Coupled Oscillators

The program in Figure 4.12 simulates a pair of similar harmonic oscillators (Chapter 2, Section 2.2.1) with

d/dt x = xdot	**d/dt xdot = - ww * x + k * y**
d/dt y = ydot	**d/dt ydot = - ww * y + k * x**

coupled by the inputs **k * y** and **k * x**. When oscillations are started by an initial displacement **x(0)** = 0.4, the coupled oscillators pass the energy back and forth as shown in Figure 4.13. Instead of programming the four derivative assignments explicitly, our program declares a submodel **oscillator** with

```
-- COUPLED OSCILLATORS
------------------------------------------------------------------
display N -18 | display C16 | display R
TMAX = 6 | DT = 0.0001 | NN = 8000 | -- timing
ww = 300
k = 50 | -- coupling coefficient
------------------------------------------------------------------
SUBMODEL oscillator(x$, xdot$, z$)
d/dt x$ = xdot$ | d/dt xdot$ = - ww * x$ + k * z$
end
---------------------
STATE x, xdot, y, ydot | -- invocation state variables
x = 0.4 | -- initial value; other 3 are 0
--
drun
-------------------------------------------------------
DYNAMIC
-------------------------------------------------------
invoke oscillator(x, xdot, y) | -- invoked twice
invoke oscillator(y, ydot, x)
-------------------------------------------------------
X = x + 0.5 | Y = y - 0.5 | -- offset display
dispt X, Y
```

FIGURE 4.12 Program demonstrating two oscillator submodels.

```
SUBMODEL oscillator(x$, xdot$, z$)
   d/dt x$ = xdot$ | d/dt xdot$ = - ww * x$ + k * z$
   end
```

and invokes it twice to model the two oscillators:

```
invoke oscillator(x, xdot, y)
invoke oscillator(y, ydot, x)
```

This neatly produces inline code for our four differential equations. It would be easy to substitute oscillators with individual stiffness and coupling parameters represented by vectors **ww$** and **k$,** or to couple additional oscillators to the two shown.

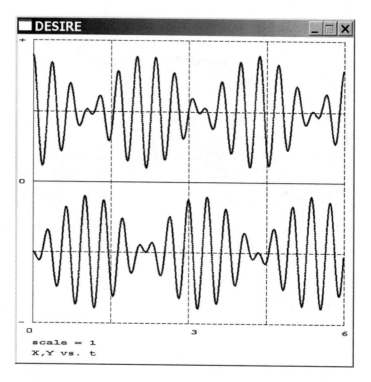

FIGURE 4.13   Solution display for the coupled-oscillator problem.

We shall return to the coupled-oscillator problem when we discuss *sub-models with vectors and matrices* in Chapter 6, Section 6.6.3.

4.6 A BANG-BANG CONTROL SYSTEM SIMULATION USING SUBMODELS

4.6.1 A Satellite Roll-Control Simulation

An interesting control system simulation will illustrate several of the new concepts discussed in this chapter. Figure 4.14 is the block diagram of a satellite roll-control system described in an Applied Dynamics International Application Report. The model uses 7 1st-order differential equations. A bang-bang controller actuates a pair of vernier roll-control rockets to make the attitude angle **thet** follow a command input **THETAC**. The resulting roll torque **f**, perturbed by an external additive torque **FD**, is integrated twice to produce the angular velocity **thetd** and the roll displacement **thet**, so that

d/dt thet = thetd | d/dt thetd = c * (f + FD)

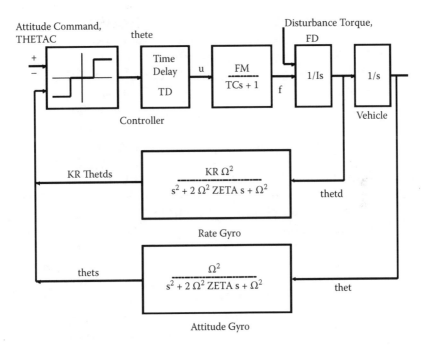

FIGURE 4.14 Simulation of a satellite attitude control system (based on an Applied Dynamics International Application Report by R.M. Howe). A 7th-order differential equation system describes vehicle and controller dynamics and two 2nd-order gyro models. The rocket controller model combines a dead-space comparator, a time delay, and a 1st-order lag that approximates valve dynamics.

where **c = 1/I**, and **I** is the satellite roll moment of inertia.

Figure 4.15 presents the complete simulation program, and Figure 4.16 is the resulting time-history display. The controller inputs are an attitude command **THETAC** and feedback from attitude and rate gyros. Both gyros are modeled by the same differential equation submodel with different invocation arguments. Each gyro submodel acts as a simple mass/spring/viscous-friction system.

For added realism, we added a small amount of hysteresis to the dead-space bang-bang control in Figure 4.14 with

thete = deadc(ethet * deltm1 - 0.01 * thete)

as in Section 4.4.3.3. As discussed in Section 4.3.5, this switching operation is preceded by a **step** statement to ensure that no switching occurs

-- R.M. HOWE'S SATELLITE ATTITUDE CONTROLLER

```
SUBMODEL gyro(x$, y$, a$, z$)
   d/dt x$ = y$   |    d/dt y$ = d * y$ + a$ * (z$ - x$)
   end
```

```
STATE thets, thetp, thetds, thetdp,f, thet, thetd
display N12   |    display C7   |    display R
A = 6 |  B = 6 |  --                      display scale factors
TMAX = 4   |    DT = 0.0025   |    NN =1000
```

```
THETAC =  0.09   | --       command pitch step, radians
DELTA1 = 0.008   | --       one-half deadspace, radians
TD = 0.025 | --                            time delay, sec
TC = 0.05  | --              actuator time constant, sec
I = 100 | --                     spacecraft inertia, kg m^2
OMEGA = 94.24478 | -- sensor natural circ. freq., 1/sec
ZETA = 0.7   | --                   sensor damping ratio
KR = 0.5  | --               rate feedback constant, sec
FM = 25  | --                actuator torque, kg m^2/sec^2
FD = 12.1875  | --    disturbance torque, kg m^2/sec^2
--
thete = 0 |   --            initialize track/hold state variable
```

```
--                                       precompute coefficients
--
a = - 1/TC   |    b = FM/TC   |    c = 1/I
deltm1 = 1/DELTA1
d = - 2 * ZETA * OMEGA   |    e = OMEGA * OMEGA
```

```
ARRAY EE[1000]   | --                            delay buffer
drun   | --                          initial values = 0
```

```
DYNAMIC
```

```
invoke gyro(thets, thetp, e, thet)   | --     attitude sensor
invoke gyro(thetds, thetdp, e, thetd)   | --     rate sensor
```

FIGURE 4.15   A complete Desire program for the satellite roll-control simulation. We added a small amount of hysteresis to the deadspace comparator.

(Continued)

```
ethet = THETAC - (thets + KR * thetds) | -- controller input
tdelay u = EE, thete, TD  |    --        actuator time delay
d/dt f = a * f + b * u  | --            actuator output torque
d/dt thet = thetd  |  d/dt thetd = c * (f + FD) | -- vehicle
  attitude
step
thete = deadc(ethet * deltm1 - 0.01 * thete)
                              | --control deadzone
```

```
theta = A * thet  |  thedot = B * thetd  | --  runtime display
THETE = 0.25 * thete - 0.5
dispt theta, thedot, THETE
```

FIGURE 4.15 (*Continued*) A complete Desire program for the satellite roll-control simulation. We added a small amount of hysteresis to the deadspace comparator.

in the middle of an integration step; **thete** is a track/hold state variable initialized by the experiment protocol.

The comparator is followed by a time delay (Chapter 6, Section 6.8.2) and a first-order lag that simulates valve dynamics.

4.6.2 Bang-Bang Control and Integration

Referring again to Section 4.3.5, the **step** statement postpones bang-bang switching until the end of the current integration step, so that realistic simulation of bang-bang control requires acceptably small integration steps **DT**. To illustrate this point, the solution displayed in Figure 4.16 shows time histories for **DT = 0.004** and **DT = 0.00004**.

Note that the slow initial transient in Section 4.5.2 does not require the expensively small **DT** values needed when there is more frequent switching later on. It would be possible to save computing time by reducing **DT** only when it is necessary. Conventional variable-step integration does not help here, since it tests only integration errors *within* integration steps. Desire's integration rules 2, 3, and 5 (Table A.1 in the Appendix), however, let *users* program **DT**, say with

$$DT = DT0 * SAT(const * abs(error)) + DTMIN$$

where **DTMIN** is a small minimum step size. **DT** becomes small whenever the absolute roll-attitude **error = thet − THETAC** is small.

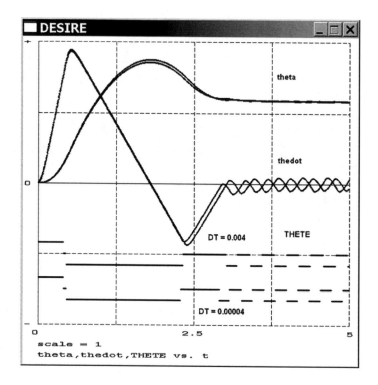

FIGURE 4.16 Stripchart-type time-history display for the satellite roll-control system. Note the effect of introducing a realistically small integration step **DT**.

This technique was first demonstrated in Reference 1 with a complete Desire program simulating the response of a bang-bang servo to a suddenly applied sine wave input. Figure 4.17 shows the step response of a bang-bang servo simulated with a similar program. The time histories show clearly how **DT** decreases for small absolute errors, and thus near each switching point.

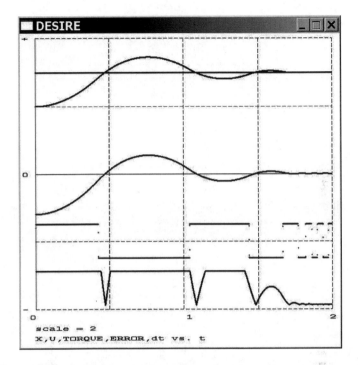

FIGURE 4.17   Simulation of bang-bang servo response to a simple input step. The program reduces the integration step size **DT** whenever the absolute value of the control system error is small.

REFERENCES

1. Korn, G.A. (2007), *Advanced Dynamic-System Simulation*, Wiley, Hoboken, NJ.
2. Cellier, F., and E. Kofman (2006), *Continuous-System Simulation*, Springer, New York.
3. Korn, G.A. (2000), Simplified Function Generators Based on Fuzzy-Logic Interpolation, *SIMPRA*, 7, 709–717.

CHAPTER 5

Programming the Experiment Protocol

5.1 INTRODUCTION

This chapter describes tools for creating effective experiment-protocol scripts. We saw in Chapter 1, Section 1.2.2, that this requires a complete programming language, much like an advanced Basic dialect. Beyond this, Desire experiment-protocol scripts can include operating-system commands prefaced by **sh**. This lets simulation experiment protocols call external programs such as spreadsheets or special graphics programs and communicate with them through files or pipes.

The following sections present the program-control and array-handling syntax used in application programs later in the text. If you wish, you can skip the remainder of this chapter until you actually require one of these language features. We shall refer to this chapter as needed.

5.2 PROGRAM CONTROL

5.2.1 Labels and Branching

A label declaration

label label2

marks a line in the experiment-protocol script with a *label* named **label2**. With **label2** declared anywhere in the script, a programmed or command-mode statement

go to label2

makes the program *branch unconditionally* to the line following the label declaration.* Listings automatically offset labels to show them more clearly. A label declaration must be the only statement on its line, except for comments.

As in other programming languages, branching into **if/else** clauses, loops, or procedures (Sections 5.2.2–5.2.4) returns an error.

5.2.2 Conditional Branching

if statements let programs *branch on conditions* such as

expression1 RELATION expression2

where **RELATION** stands for one of the operators **<, >, =, <=, >=, <> (<>** means "not equal"),† as in

if Crit >= 0 then q19 = x * 1000 | else proceed

then and/or **else** clauses can have multiple statements:

```
if A21 < 0 then p = p + 1 | q = q + 50
   else p = p - 1 | write 'exception!'
   proceed
```

Code between **then** and the corresponding **else** executes if and only if the specified condition is true. Otherwise the program continues on the far side of **else. proceed** statements simply let the program continue; they neatly delimit *nested* **if** *clauses*. There must always be exactly one **else** and one **proceed** for each **if** in the script. This syntax makes our small experiment-protocol interpreter very efficient.

In *nested* **if** statements, the outermost **if** corresponds to the outermost **else** that is still free. The **else** clause terminates with the corresponding outermost **proceed**. Listings automatically indent ("prettyprint") **if** clauses to sort out nested statements.

---

* The Reference Manual on the book CD also describes branching to Desire line numbers, and computed go to.
† Desire **if** statements also admit two-term logical expressions *condition1* **or** *condition2* and *condition1* **and** *condition2*. The same is true for the **while** and **until** conditions in Section 5.2.3.

When an **if ... then ...** construction does not fit one line, you can use a comment delimiter (**--** , two hyphens) as the first statement of the **then** clause.

5.2.3 **for, while,** and **repeat** Loops

Our experiment-protocol scripts admit Pascal-like *program loops*:

```
for y = 0 to PI/4 step PI/40
   sinus = sin(y)
   next
k=1
while k <= 80
   write k, k * k, 1/k
   k = k + 1
   end while
k = 0
repeat | k = k + 10 | write k, k * k, 1/k | until k > 100
```

Listings automatically indent loops as shown. An **exit** statement in a loop causes a goto-less exit to the line following the current loop. **exit** may be followed only by comments, **else** clauses, or **proceed** on the same line. Listings automatically indent (prettyprint) loops.

In *command mode* (i.e., as a typed command), Desire admits only *one-line* loops such as

for i = 1 to 10 | a = 2 * i | write a | next

and displays a warning message.

Loops may be *nested* but as in other programming languages, must never overlap other loops or **if/then/else** clauses. There must be a unique **else, proceed, next, end while,** and **until** corresponding to each **if, else, for, while,** and **repeat**. The program returns error messages on loop overlaps and on duplicate **next, end while,** and **until** statements if the program actually reaches the faulty code.

5.2.4 Experiment-Protocol Procedures

Experiment-protocol *procedures* are declared in the form

```
PROCEDURE procname(x, y, ...; X, Y, ...)
... procedure code involving x, y, ...,X,Y, ..., other ("global")
    variables
end
```

Listings automatically indent the procedure body. The semicolon separates dummy arguments **x, y, ...** which will be called *by value* (typically expressions serving as procedure *inputs*) and dummy arguments **X, Y, ...** that will be called *by name* (like **VAR** parameters in Pascal; they are usually procedure *outputs*). Here is an example:

```
PROCEDURE xform(r, theta; X, Y)
    X = r * cos(theta)
    Y = r * sin(theta)
    end
```

Procedures must not contain **label** declarations. User-defined procedures may have only value parameters, only **VAR** parameters, or no arguments at all:

```
PROCEDURE foo(p, q)
PROCEDURE fee(; g, h, i)
PROCEDURE fum()
```

Redundant semicolons and parentheses must not be omitted. Desire "protects" dummy arguments by returning an error message when you try to reuse their names.

An **exit** statement anywhere within a procedure causes an immediate return.

Once defined, a procedure can be *called* later in the experiment protocol (but not in DYNAMIC program segments) with

```
call procname(a, b, ...; A, B, ...)
```

You can substitute expressions for value parameters **a, b, ...** . The **VAR** parameters **A, B, ...** can be named variables or arrays, but not subscripted array elements. There can be multiple calls, with the same or different arguments, to the same procedure. The procedure **xform** in our example could be called with

call xform(r1, theta1; x1, 1)

and

call xform(RADIUS - 2.12, 4 * PI/180; xx, y3)

The program returns an error when arguments and/or argument types in a procedure declaration and call do not match.

Procedure calls can be nested; but a procedure must not call itself directly or indirectly. **PROCEDURE** definitions can contain procedure and function calls, but no declarations. Recursive **PROCEDURE** definitions are illegal.

5.3 ARRAYS AND SUBSCRIPTED VARIABLES

5.3.1 Arrays, Vectors, and Matrices

5.3.1.1 Simple Array Declarations

Many simulation models employ one- and two-dimensional real *arrays* representing time histories, vectors and rectangular matrices.* The experiment-protocol script must *declare* each array before it or its elements (*subscripted variables*) can be used in the experiment protocol or in DYNAMIC program segments.

A one-dimensional array (*vector*) **v** of real subscripted variables **v[1]**, **v[2], …, v[n]** is declared with

ARRAY v[n]

A two-dimensional array (*matrix*) **W** of real subscripted variables **W[1, 1]**, **W[1, 2]**, … , **W[m, n]** is similarly declared with

ARRAY W[m,n]

The matrix has **m** *rows and* **n** *columns:*

W[1,1] W[1,2] … W[1,n]
W[2,1] W[2,2] … W[2,n]
....
W[m,1] W[m,2] … W[m,n]

---

* Desire also admits arrays with more than two dimensions and arrays of integers and complex numbers (Chapter 8, Section 8.3.1). Such arrays are rarely used.

The array dimensions **m, n** are positive real expressions truncated to integer values.

A declaration line can define more than one array, as in

ARRAY v[3], y[12], A[m,n], B[120,7], v[100], ...

Multiple declaration lines are legal. All array elements (*subscripted variables*) initially default to 0. The experiment protocol may fill arrays with assignments to subscripted variables, with **data/read** assignments, or through console or file input (Sections 5.3.2 and 5.3.4).

The programmed or command-mode statement **clear** clears all array definitions (and also definitions of variables, labels, user-defined functions, and procedures) and reclaims their memory allocations. A **clear** statement must precede any attempt to redimension an array.

DYNAMIC program segments as well as experiment-protocol scripts can manipulate one- and two-dimensional arrays as *vectors* and *matrices*. Vector/matrix operations will be introduced in Chapter 6.

5.3.1.2 Equivalent Arrays

The modified array declaration

$$\text{ARRAY v1[n1] + v2[n2] + ... = v} \tag{5.1a}$$

creates a set of *subvectors* **v1, v2, ...** concatenated into a vector **v** of dimension **n1 + n2 + ...** . The elements **v[1], v[2], ...** of **v** overlay the subvectors **v1, v2, ...** , starting with **v1**, so that **v1[3]** is equivalent to **v[3]**, **v2[2]** is equivalent to **v[n1 + 2]**, etc. We have already used this technique in Chapter 4, Section 4.2.2, to combine abscissa and ordinate breakpoint arrays for table-lookup function generators. The most important application of subvectors, though, is neural network simulation [1].

A second modified array declaration

$$\text{ARRAY W[m,n] = v} \tag{5.1b}$$

declares the same array both as a two-dimensional array (matrix) **W** and as a one-dimensional array (vector) **v** of dimension **mn**, with

$$W[i, k] = v[m * (i - 1) + k] \quad (i = 1, 2, ...,m; k = 1, 2, ..., n)$$

This enables very powerful matrix operations (Chapter 6) and is also used to implement fuzzy-set logic [1]. The short program

```
m = 3 | n = 4
ARRAY W[m,n] = v
for i = 1 to m | for k = 1 to n
  W[i,k] = (i - 1) * n + k
  next | next
write W, v
```

demonstrates the equivalence of **W** and **v**. Both array declarations (Equation 5.1) act like equivalence statements in Fortran.

5.3.1.3 *STATE* Arrays
As we saw in the preceding chapters, ordinary scalar differential equation state variables need not be declared before they are used. But *subscripted state variables* **x[1]**, **x[2]**, ..., **y[1]**, **y[2]** used in DYNAMIC-segment derivative assignments such as

$$d/dt\ x[13] = ... \qquad d/dt\ y[2] = ...$$

and/or implied by vector differential equations (Chapter 6, Section 6.3.1) require experiment-protocol declarations such as

STATE x[n]   **STATE x[n], y[m], ...**

that define one-dimensional arrays (*state vectors*) **x**, **y**, ...* Initial values of all state variables are set by the experiment protocol or default to 0.

5.3.2 Filling Arrays with Data
5.3.2.1 Simple Assignments
Experiment-protocol scripts can fill previously declared arrays with assignments to subscripted variables, as in

```
A[19,4] = 7.3 | v[2] = a – 3 * b
for i = 1 to n | x[i] = 20 * i | next
```

or from **data** lists and **read** assignments.

---

* **STATE** declarations are also needed for state variables defined by submodels (Chapter 4, Section 4.5.1.3).

5.3.2.2 **data** *Lists and* **read** *Assignments*
data *lists* such as

data expression1, expression2, ...

supply numerical values for **read** assignments such as

read *variable name*
read *array name*
read *variable or array name, variable or array name, ...*

Successive **read** statements assign values of successive **data** expressions to the variables in the **read** list. The system *read pointer* normally starts at the first expression of the first **data** list and searches for the next **data** list as each **data** list is exhausted. An error message indicates insufficient data.

As an example,

x = 117.222
ata 11, 12, 13, 14
y = 100. 222
data 15, 16, x - y, 18
read x1, x2, x3, x4
read x5, x6, x7, x8

assigns

x1 = 11, x2 = 12, x3 = 13, x4 = 14, x5 = 15, x6 = 16, x7 = 7, x8 = 18

read operations read two-dimensional arrays in row-major sequence. For convenience in separating data for different arrays, or to indicate array rows or columns, one can substitute semicolons for commas anywhere in a **data** or **read** list. Thus:

ARRAY Amatrix[3,3]
x = 27.5 | y = - 20.6
data 200.19 ; 10, 20,30 ; 40, 50, 60 ; x + y, 80, 90
read velocity, Amatrix

produces velocity = 200.19 and

Amatrix[1,1] = 10   Amatrix[1,2] = 20   Amatrix[1,3] = 30
Amatrix[2,1] = 40   Amatrix[2,2] = 50   Amatrix[2,3] = 60
Amatrix[1,3] = 6.9 Amatrix[2,3] = 80   Amatrix[3,3] = 90

The programmed or command-mode statement **restore** *resets the data pointer* to the first expression in the first **data** list. Desire also admits **restore** *labelname*, which resets the data pointer to the first **data** item following the program line referenced by a label (Section 5.2.1).

The data pointer is automatically reset to the first **data** expression in the script whenever a program line is modified (as during debugging).

5.3.2.3 Text-File Input
Section 5.4.2.2 shows how **input** is used *to fill an array from a text file*.

5.4 EXPERIMENT-PROTOCOL OUTPUT AND INPUT

5.4.1 Console, Text-File, and Device Output
5.4.1.1 Console Output
Programmed or command-mode **write** statements display, print, or store the results of computations. For example:

xvar = 222.22 * 3 - 4.7
write xvar

or

write 222.22 * 3 - 4.7

evaluates an expression and displays the result in the Command Window. Programmed or command-mode statements such as

write 'The moon is pink'

display text strings. As in Basic, one can write multiple items on a line, for example,

write 'A = ';A,'B = ';B,'C = ';C

Also as in Basic, a semicolon produces the next item immediately following the last one, while a comma inserts a tab between substrings. Completion of each **write** statement causes a carriage return/line feed (newline) unless the **write** statement ends with a semicolon. **write** alone causes a newline **write ;** (note the blank) would return an error message.

Desire **write** statements correctly recognize different previously declared data types including integers, complex numbers, and arrays and produce the correct output, as in

AA= - 231.77
ARRAY vector1[800]
....
write AA, vector1

5.4.1.2 File and Device Output

To write to an ASCII data file or device (serial printer, data link), one must first **connect** (open) a "channel" (really a buffer in memory) for the file or device. Programmed or command-mode statements such as

connect 'BRIE.dat' as output 4
connect 'lpt2' as output 5

associate named files or devices with *channel numbers* between 0 and 10. Double quotes may replace single quotes. Once a file or device is **connect**ed, one can write to it with a command-mode or programmed **write** statement referencing the numbered channel:

write #4,'x14 has the value ';B

A comma must follow the channel number. There can be more than one such line. File extensions for **connect** statements default to **.dat**.

Desire **write** statements recognize previously declared array names and label each array with a header listing its type, name, and dimension or dimensions. This is nice for displays and printed listings, but files with such labels are not machine-readable. For this reason, Desire also provides modified **write** statements

write ##4, a, b, array1, . . .

(note the double **##**), which simply write successive scalars and/or array elements separated by tabs. This produces machine-readable text files without headers. Such files can, for instance, feed spreadsheet or database programs.

5.4.1.3 Closing Files or Devices
When you are finished with your file or device, **disconnect** (close) its channel with a statement such as

disconnect 3
or **disconnect 3, 4, . . .** (disconnects multiple channels)

Each **disconnect** operation flushes data remaining in a buffer or buffers, releases the channel number and, for file output, makes the proper mass-storage directory entry.* The current default device or path is implied if no device or path is specified.

5.4.2 Console, File, and Device Input
5.4.2.1 Interactive Console Input
The programmed or command-mode statement

input x, y, z, . . .

produces successive Command Window prompts such as **x?** asking the user to type a value or expression for **x**, **y**, … . The program then assigns the resulting numbers to **x**, **y**, **z**, … . Any error produces an error message and a new prompt for the same item. You can *abort* the **input** operation by typing **!**.

If **X** is a (previously declared) *array*, then **input X** produces successive prompts **X?** until the array is filled. These prompts do not specify the array index.

5.4.2.2 File or Device Input
After opening a device or text file for input with

connect 'GRUNT.dat' as input 3

a programmed or command-mode statement

---

* **STOP** and **new** statements do not automatically **disconnect** (close) open files. But any error, or exit from Desire, automatically closes all open files and thus writes output files.

input {#*channel number*,} *variable* {,*variable*, ...}

automatically reads successive values or expressions for each variable from the specified channel into each variable. File entries must be separated by line feeds, commas, or semicolons. Line-feed-delimited files must be terminated with a line feed. The program returns an error message if end-of-file (EOF) is reached.

Previously declared **arrays gee**, **vee**, ... can be filled from text files* with programmed or command-mode statements such as

input #3, gee
for i = 1 to 10 | input #3, gee[i] | next

or

input #3, gee, vee, ...

Files or devices connected for input must be closed after use, as in

disconnect 3

5.4.3 Multiple DYNAMIC Program Segments

Most experiment protocols call simulation runs that exercise a model defined in a single DYNAMIC program segment. But Desire simulation studies can also *run and compare different models, different versions of the same model, and/or different displays* without having to start a new program.

Desire admits multiple DYNAMIC program segments following the DYNAMIC line in any order. Besides the usual unlabeled DYNAMIC segment (which can also be omitted), there may be any number of *labeled DYNAMIC segments*, each beginning with a user-named label declaration

label *labelname*

---

* Note that different **c** systems may have different end-of-file (EOF) markers. If this causes problems, pad data files with extra 0 entries. The example programs **input.src, output*.src** on the book CD demonstrate different cases.

As in Section 5.2.1, only comments may follow a label declaration on the same line. The experiment protocol can compile and run labeled DYNAMIC segments in any desired order with

drun *labelname* or drunr *labelname*

To save time, **drun** or **drunr** reruns or continues unlabeled DYNAMIC segments without recompilation. Labeled DYNAMIC segments, on the other hand, are compiled every time they run; the resulting simulation runs cannot be continued.

Different DYNAMIC segments can reference and modify the same and/or different variables and parameters.* They can also exchange time-history arrays with **store** and **get** statements (Chapter 6, Section 6.8.1).

Multiple DYNAMIC segments can run repeatedly, one at a time, in any desired sequence. They can represent different models, or different versions of a model. But they often just process, display, or print data before or after a true simulation run. A simulation might, for example, **plot x** and **y** versus **t** and then run a labeled DYNAMIC program segment that does nothing but plot **y** versus **x**. Sections 6.4.3.1 (Chapter 6), 7.4.1 (Chapter 7), and 8.2.2 (Chapter 8) show examples.

5.5 EXPERIMENT-PROTOCOL DEBUGGING, NOTEBOOK FILE, AND HELP FILES

5.5.1 Interactive Error Correction

The Command Window reports programming and runtime errors with an error message and usually displays a guilty program line with its Desire line number. One can then edit the program and run again.

As a precaution, you might want to preserve the current Editor Window and type **ed** to produce a new Editor Window for correcting the error. Use **edit** instead of **ed** if you want to refer to line numbers. **keep** or **keep** *'filespec'* preserves the corrected file.

---

* In the unlikely case that a derivative assignment **d/dt x = ...** for the same state variable x is used in more than one DYNAMIC program segment, the experiment-protocol script must declare **x** with a **STATE** declaration **STATE x, ...**, like those for subscripted state variables (Chapter 6, Section 6.3.1).

5.5.2 Debugging Experiment-Protocol Scripts

Desire's script-debugging facilities can stop or single-step the experiment protocol and trace program operation:

1. Programmed **STOP** statements inserted between script lines act as *breakpoint*s that stop execution. One can then examine variables with command-mode or programmed **write** or **dump** statements (see below) and restart execution by typing **go**.

2. The typed command **trace** *toggles single-stepping* of the experiment-protocol script. The program responds with **trace ON** or **trace OFF**. Type **run** to start, and then **go** to continue after each line. At each step, the program will display the current program line and report all assignments, and also executions of **FUNCTION**, **PROCEDURE**, and **go to**.

3. The commands **trace, rld, reload, new,** and **NEW** stop single-stepping.

4. The command **dump** displays all currently used variables, including system variables such as **NN, TMAX**, etc... .* **dump #n** dumps to a file or device previously **connect**ed to Channel **n** (Section 5.4.1.2).

You can **write**, modify, or **dump** variables while single-stepping. The program may, of course, not continue correctly on **go** if you add or delete program lines.

5.5.3 The Notebook File

When Desire starts, the program automatically opens a *notebook file* (journal file, log file) **NOTES.JRN** in the **mydesire** folder. This notebook file automatically records

1. The *date and time* when the program is first loaded, and whenever a programmed or command-mode **time** statement executes.

2. The *problem identification code* (file specification) of the current program whenever a programmed or command-mode **PIC, note, reload, keep,** or **keep+** statement executes.

---

* Under Linux one can stop this possibly long listing with **ctrl-c**.

One can also record user comments and selected program lines (see the Reference Manual), but it is more convenient to enter comments in extra Editor Windows produced with **ed**. Editor Windows can be saved as needed.

5.5.4 Help Facilities

Desire help screens are ordinary *text files created by users*. The help text can be dragged to an Editor Window for display, or it can be displayed with any other screen editor.

Help files can be short instructions, manual excerpts, or all kinds of messages. The Desire distribution includes only a few examples in the **mydesire** folder and in a subfolder **HELP**.

The programmed or command-mode statement

help *filename*

lists any text file *filename* stored in the **mydesire** folder in the Command Window; typing **space** continues long text files. This is one way to display user instructions in the course of a simulation study. Experiment-protocol scripts can, instead, use operating-system commands prefaced with **sh** (Section 5.1.1) to call external programs producing any desired text and/or graphics.

REFERENCE

1. Korn, G.A. (2007), *Advanced Dynamic-System Simulation: Model-Replication Techniques and Monte Carlo Simulation*, Wiley, Hoboken, NJ.

CHAPTER **6**

Models Using Vectors and Matrices

6.1 OVERVIEW

6.1.1 Introduction

In Chapter 5, Section 5.3.1, we declared an **n**-dimensional *vector* **v** with **ARRAY v[n]**, and an **n**-by-**m** *matrix** **W** with **ARRAY W[n,m]**. We showed how such arrays are "filled" with subscripted variables **v[i]** and **W[i, k]** that initially default to 0. Vectors and matrices, similar to the submodels in Chapter 4, Sections 4.5.1, 4.5.2, and 4.6.1, are not simply mathematical shorthand; they normally represent intuitively meaningful model objects. Thus, vector arrays can model physical vectors (e.g., displacements, velocities, forces), arrays of time history samples, and neuron layer activations. Examples of rectangular matrices are rotation, state-transition, feedback, and connection-weight matrices. Desire experiment protocols and DYNAMIC program segments permit efficient vector/matrix operations programmed in a readable mathematical notation.

We begin with simple vector and matrix assignments that work only in experiment-protocol scripts (Sections 6.2.1–6.2.5) and then turn to runtime-compiled DYNAMIC-segment vector and matrix operations. DYNAMIC program segments handle vector/matrix functions, differential equations, and difference equations as easily as their scalar counterparts.

---

* An **n**-by-**m** matrix declared with **ARRAY W[n,m]** has **n** rows and **m** columns.

Desire's compiler can replicate a dynamic-system model with different parameters and simulate 100 to 10,000 such models in *one* simulation run. This permits very efficient parameter-influence and statistical studies.

6.2 VECTORS AND MATRICES IN EXPERIMENT-PROTOCOL SCRIPTS

6.2.1 Null Matrices and Identity Matrices

A *square matrix* declared with **ARRAY A[n,n]** is by default a *null matrix* with all **A[i, k]** equal to zero (Chapter 5, Section 5.3.1). One can reset any existing square matrix **A** to a null matrix with the matrix assignment

MATRIX A = 0

The assignment

MATRIX A = 1

turns a previously declared square matrix **A** into an **n**-by-**n** *identity matrix* with diagonal elements equal to 1 and all other elements equal to 0. An error message results if **A** is not square.

As an example, the script

```
N = 2 | m = 3
ARRAY A[n,n], B[m, m]
MATRIX A = 0
MATRIX B = 1
write A, B
```

produces

```
A[2,2] : REAL ARRAY (in ROW major sequence)
0.000000e+000 0.000000e+000
0.000000e+000 0.000000e+000

B[3,3] : REAL ARRAY (in ROW major sequence)
1.000000e+000 0.000000e+000 0.000000e+000
0.000000e+000 1.000000e+000 0.000000e+000
0.000000e+000 0.000000e+000 1.000000e+000
```

6.2.2 Matrix Transposition

Given a previously declared and filled **n**-by-**m** matrix **A** and a previously declared **m**-by-**n** matrix **B**, the assignment

MATRIX B = A%

turns **B** into the *transpose* **B** of **A**, with **B[i,k] = A[k, i]** for all **i** and **k**. An error message is returned if **B** is not an **m**-by-**n** matrix.

6.2.3 Matrix/Vector Sums and Products

Experiment-protocol programs admit vector assignments of the form

Vector v = alpha * x + A * y + . . .

where **v**, **x**, and **y** are vectors, and **A** is a rectangular matrix. **alpha** is a scalar, which can be a literal number but not an expression. An error message results if the vectors and matrices are not conformable, and also if the vector **x** in a matrix-vector product such as **A * x** is identical with the assignment target (that would cause an illegal recursion, as in Section 6.3.2.1). Experiment-protocol scripts do not accept the more general vector expressions discussed in Section 6.3.1.

Script **DOT** *statements* produce inner products such as

DOT beta = a * b DOT sum = x * 1 DOT gamma = x * A * b

and also sums of such products, similar to **DOT** assignments in DYNA-MIC program segments (Section 6.5.1).

6.2.4 **MATRIX** Products

The experiment-protocol matrix assignment

MATRIX D = alpha * A * B * C * . . .

implements fast multiplication of previously declared and filled *square* matrices **A**, **B**, **C**, all with the same dimensions. The optional first term **alpha** is a scalar multiplier, which can be a literal number but not an expression.

6.2.5 Matrix Inversion and Solution of Linear Equations

Given two previously declared square matrices **A**, **B** with equal dimensions, the script matrix assignment

MATRIX B = $In(A)
MA\TRIX B = $In(A)

produces the *matrix inverse* **B** = **A**$^{-1}$ of **A**, or an error message if the inverse does not exist. The short script in Figure 6.1 defines a nonsingular square matrix **A** and a vector **b**, finds the matrix inverse **A**$^{-1}$, and produces the *solution*

$$x = A^{-1} b$$

of the system of linear equations

$$A x = b$$

The program then tests this result by computing the error vector **error = A x - b** obtained on substituting the solution vector **x** in the given equations. One can also test the quality of the matrix inversion by reinverting the computed inverse matrix.

6.3 VECTORS AND MATRICES IN DYNAMIC-SYSTEM MODELS

6.3.1 Vector Expressions in DYNAMIC Program Segments

Vectors and matrices declared in the experiment protocol can be used in DYNAMIC program segments. At this point, we recall that arrays **x** of subscripted *differential equation state variables* **x[1]**, **x[2]**, ... (state vectors) must be declared with **STATE x[n]** (Chapter 5, Section 5.3.1.3).

Desire's DYNAMIC-segment vector assignments are far more powerful than simple vector algebra because they can use very general *vector expressions*. With vectors **x**, **y**, **z**, ... all of equal dimension **n**

Vector y = f(t; x, z, ...) compiles into
y[i] = f(t; x[i], z[i], ...) (i =1,2, ..., n)

Vectr d/dt x = f(t; x,y,...) compiles into
d/dt x[i] = f(t; x[i], y[i], ...) (i =1,2, ..., n)　　　　(6.1)

f() stands for any expression that can be used in a scalar assignment and may involve literal numbers, scalar parameters, parentheses, and functions—even table-lookup and user-defined functions. An error is returned when you try to combine vectors with unequal dimensions.

Each vector assignment automatically generates **n** *scalar assignments.* The resulting code is fast, for the vectorizing compiler eliminates all run-time vector-loop overhead.

For example, given **n**-dimensional vectors **y, u, v,** and **z**

Vector y = (1 − v) * (cos(alpha * z * t) + 3 * u)

compiles into the **n** assignments

y[i] = (1 - v[i]) * (cos(alpha * z[i] * t) + 3 * u[i])   (i = 1, 2, ..., n)

DYNAMIC program segments can freely combine scalar assignments and multiple vector assignments with the same or different dimensions. Vector assignments can, moreover, accept *matrix/vector products* (Sections 6.3.2 and 6.3.3) *and index-shifted vectors* (Section 6.4.1).*

6.3.2 Matrix/Vector Products in Vector Expressions

6.3.2.1 Matrix/Vector Products
Any of the **n**-dimensional vectors in a vector expression (Equation 6.1), say **z**, can be a *matrix-vector product* **A * v**, where **A** is a constant or variable **n**-by-**m** matrix, and **v** is an **m**-dimensional vector.† Thus

Vector y = tanh(A * x)

automatically compiles into the **n** *scalar assignments*

$$y[i] = \tanh(\sum_{k=1}^{m} A[i,k] * v[k]) \qquad (i = 1, 2, ... , n)$$

---

* The Desire Reference Manual on the book CD describes additional matrix/vector operations used mainly in neural-network simulations [3].
† **A** and **v** must be defined by earlier assignments; they cannot be expressions.

In matrix/vector products written as **A%** * **x**, Desire automatically *transposes* the given matrix **A**. Thus, if **A** is an **m**-by-**n** matrix and **v** is an **m**-dimensional vector, the vector assignment

Vector q = exp(A% * v)

compiles into the **n** scalar assignments

$$q[i] = \exp\left(\sum_{k=1}^{m} A[k,i] * v[k]\right) \qquad (i = 1, 2, \ldots, n)$$

Nonconformable matrices are automatically recognized and cause error messages.

6.3.2.2 Example: Rotation Matrices

The rotation of a plane vector **x** ≡ **(x[1], x[2])** into the vector **y** ≡ **(y[1], y[2])** can be programmed with two scalar defined-variable assignments:

y[1] = x[1] * cos(theta) - x[2] * sin(theta)
y[2] = x[1] * sin(theta) + x[2] * cos(theta)

where the variable **theta** is the rotation angle. One can, instead, declare the *two-dimensional rotation matrix* **A** with **ARRAY A[2,2]** and then specify the time-variable elements **A[i,k]** of **A** in a DYNAMIC program segment:

A[1,1] = cos(theta) | A[1,2] = - sin(theta)
A[2,1] = - A[1,2] | A[2,2] = A[1,1]
VECTOR y = A * x

The matrix/vector product **A** * **x** represents the rotated vector **y** compactly and can be used in vector expressions (Figure 6.1).

A rotation matrix is especially useful when one wants to rotate not just one vector, but several different vectors **x**, **p**, **q**, … through the same angle **theta**. The rotated vectors are **A** * **x**, **A** * **p**, **A** * **q**, … . Analogous three-dimensional rotation matrices are used, for instance, in flight simulations.

-- MATRIX INVERSION AND LINEAR EQUATIONS

ARRAY A[4, 4], Ainverse[4, 4], C[4, 4]
ARRAY x[4], b[4], error[4]
data 1, 3, 50, 3; 2, 3, 0, -1; 0, -1, 2, 0; -90, 0, 1, 3 | read A
data 2, 3, 5, -7 | read b
--
MATRIX Ainverse = $In(A) | write Ainverse
write "type go to continue" | STOP
Vector x = Ainverse * b
Vector error = A * x - b
write x, error

FIGURE 6.1   This short script produces the inverse **Ainverse** of a square matrix **A**, solves the set of linear equations **Aww x = b**, and determines the solution errors.

6.3.3 Matrix/Vector Models of Linear Systems

In Chapter 4, Section 4.5.2, we modeled a system of two coupled harmonic oscillators with

d/dt x = xdot   d/dt xdot = - ww * x + k– x
d/dt y = ydot   d/dt ydot = - ww * y + k * y

(Figure 6.2a). These four scalar assignments can be replaced with *a single vector assignment*

$$
\text{Vectr } d/dt \; x = A * x \text{ with } A = \begin{Vmatrix} 0 & 1 & 0 & 0 \\ -ww & 0 & k & 0 \\ 0 & 0 & 0 & 1 \\ k & 0 & -ww & 0 \end{Vmatrix}
$$

(Figure 6.2b), or alternatively with *two vector assignments*, one for each oscillator,

$$
\begin{array}{lll}
\text{Vectr } d/dt \; x = A * x + B * y & & \\
& A = \begin{Vmatrix} 0 & 1 \\ -ww & 0 \end{Vmatrix} & B = \begin{Vmatrix} 0 & 0 \\ k & 0 \end{Vmatrix} \\
\text{Vectr } d/dt \; y = A * y + B * x & &
\end{array}
$$

(as shown in Figure 6.2c).

```
--                     COUPLED OSCILLATORS
----------------------------------------------
display N11 | display C7
TMAX = 10 |   DT = 0.0001|   NN = 10001
ww = 300 |     k = 50
```

(a)
```
x = 0.4 | drun | –               set initial value and run
------------------------------------------------------------
DYNAMIC
------------------------------------------------------------
d/dt x = xdot |   d/dt xdot = - ww * x + k * y
d/dt y = ydot |   d/dt ydot = - ww * y + k * x
```

(b)
```
STATE x[4]   |   ARRAY A[4, 4]
data 0, 1, 0, 0; - ww, 0, k, 0; 0, 0, 0,1; k, 0, - ww, 0
read A
x[1] = 0.4 | drun | --           set initial value and run
------------------------------------------------------------
DYNAMIC
------------------------------------------------------------
Vectr d/dt x = A * x
```

(c)
```
STATE x[2], y[2]   |   ARRAY A[2, 2], B[2, 2]
data 0, 1; - ww, 0 |   read A
data 0, 0; k, 0 |   read B
x[1] = 0.4 | drun | --           set initial value and run
------------------------------------------------------------
DYNAMIC
------------------------------------------------------------
Vectr d/dt x = A * x + B * y
Vectr d/dt y = A * y + B * x
```

FIGURE 6.2 Programs simulating a pair of coupled oscillators: (a) Using four scalar differential equations; (b) Using a single vector differential equation; (c) Using two vector differential equations, one for each oscillator. The three program lines at the top are common to all three programs.

Our simple example is a special case of the *general matrix/vector model of a linear dynamic system* used in control engineering [1],

Vectr d/dt x = A * x + B * u   Vector v = C * x + D * u

where **x** is an **n**-dimensional *state vector,* **u** is an **m**-dimensional *input-signal vector,* and **v** is an **N**-dimensional *output vector.* The matrices **A, B, C, D** defining the dynamic system can be functions of the time **t**.

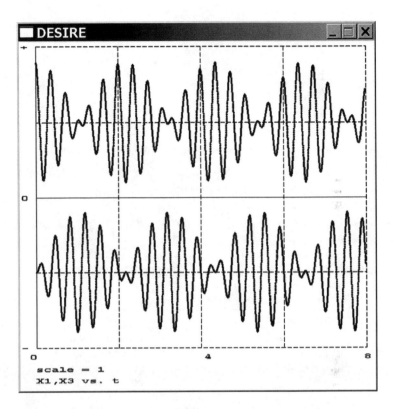

FIGURE 6.2d   Time histories for the coupled-oscillator simulation.

6.4 VECTOR INDEX-SHIFT OPERATIONS

6.4.1 Index-Shifted Vectors

Given an **n**-dimensional **[1], x[2], ..., x[n])** and a constant positive or negative integer **k**, the *index-shifted vector* **x{k}** is the **n**-dimensional vector **(x[1+k], x[2+k], ..., x[n+k])**. Index-shifted-vector components with indices less than 1 or greater than **n** are simply set to 0.

Significantly, the program line

Vector x = x{-1} | x[1] = input

compiles into a system of difference equations

x[i] = x[i - 1] (i = 1, 2, ..., n) x[1] = input

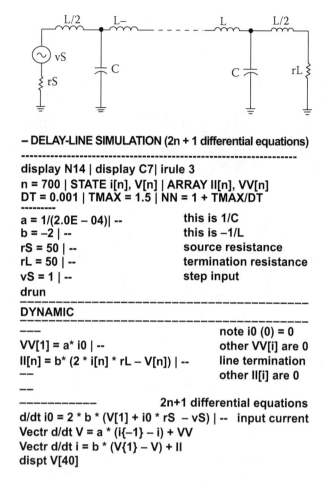

```
– DELAY-LINE SIMULATION (2n + 1 differential equations)
-----------------------------------------------------------------
display N14 | display C7| irule 3
n = 700 | STATE i[n], V[n] | ARRAY II[n], VV[n]
DT = 0.001 | TMAX = 1.5 | NN = 1 + TMAX/DT
---------
a = 1/(2.0E – 04)| --          this is 1/C
b = –2 | --                    this is –1/L
rS = 50 | --                   source resistance
rL = 50 | --                   termination resistance
vS = 1 | --                    step input
drun
------------------------------------------------------
DYNAMIC
------------------------------------------------------
---                            note i0 (0) = 0
VV[1] = a* i0 | --             other VV[i] are 0
II[n] = b* (2 * i[n] * rL – V[n]) | --   line termination
--                             other II[i] are 0

--

------------         2n+1 differential equations
d/dt i0 = 2 * b * (V[1] + i0 * rS  – vS) | --  input current
Vectr d/dt V = a * (i{–1} – i) + VV
Vectr d/dt i = b * (V{1} – V) + II
dispt V[40]
```

FIGURE 6.3   Simulation of an inductor/capacitor delay-line circuit using Desire vector/matrix operations.

This neatly models shifting successive samples of a function **input(t)** into a tapped delay line with tap outputs **x[1] = input, x[2], ..., x[n]**. Note that **x[1] = input** overwrites the **Vector** operand and was originally developed to simplify programming neural networks and autoregressive moving-average (ARMA) control system models [1–3]. Other applications include fuzzy logic and partial differential equations [3]. The following two sections show how index shifting can reduce complicated simulation programs to just a few lines.

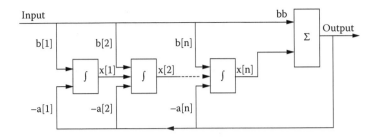

FIGURE 6.4 Block diagram of a linear dynamic system with the transfer function $H(s) = (bb\ s^n + b[n]s^{n-1} + b[n-1]s^{n-2} + \dots + b[1])/\{s^n + a[n]s^{n-1} + a[n-1]s^{n-2} + \dots + a[1]\}$.

6.4.2 Simulation of an Inductance/Capacitance Delay Line

The electrical delay line circuit in Chapter 2, Section 2.2.4, had **n + 1** = 701 inductors and **n** = 700 capacitors so that the model requires 1401 differential equations (Chapter 2, Figure 2.6). In Figure 6.3, *two index-shift operations replace 1400 derivative assignments.* Both programs produce identical results (Chapter 2, Figure 2.7).

6.4.3 Programming Linear-System Transfer Functions

6.4.3.1 Analog Systems

The block diagram in Figure 6.4 represents a classical linear dynamic system with the transfer function [1]

$$H(s) = (bb\ s^n + b[n]s^{n-1} + b[n-1]s^{n-2} + \dots + b[1])/\{s^n + a[n]s^{n-1} + a[n-1]s^{n-2} + \dots + a[1]\}$$

$$(6.2)$$

This representation applies, in particular, to analog filters. The coefficients **a[i]** and **b[k]** are often constant, but they can also be functions of time. The block diagram's chain of integrators implement an **n**th-order differential equation system that is easily modeled with the DYNAMIC-segment lines:

$$\text{input} = (\textit{a given function of the time variable } t)$$
$$\text{output} = x[n] + bb * \text{input}$$
$$d/dt\ x[1] = b[1] * \text{input} - a[1] * \text{output}$$
$$d/dt\ x[2] = x[1] + b[2] * \text{input} - a[2] * \text{output}$$
$$\dots\dots\dots\dots\dots$$
$$d/dt\ x[n] = x[n-1] + b[n] * \text{input} - a[n] * \text{output} \qquad (6.3)$$

Execution starts with given initial values for each state variable **x[i]** on the right-hand side. The initial values **t0** and **x[i]** usually default to zero.

But there is a much better way. Instead of programming **n + 2** scalar assignments, we declare **n**-dimensional vectors **x**, **a**, and **b** with

STATE x[n] | ARRAY a[n], b[n]

and program the complete linear system model, for any order **n**, *in only three lines*:

input = (*given function of* t)
output = x[n] + bb * input
Vectr d/dt x = x{-1} + b * input - a * output

These three lines compile automatically into the assignments (Equation 6.3); note how the index-shift operation models the chain of cascaded integrators. There is, once again, no runtime vector-loop overhead.

To produce the *impulse response* of such a system one programs **input = 0** and sets the initial value of **x[1]** to 1. The amplitude/phase frequency response is then obtained with Desire's FFT routine. Figure 6.5 illustrates the simulation of a simple bandpass filter.

6.4.3.2 Digital Filters

The block diagram in Figure 6.6 represents a linear sampled-data system or digital filter with the z transfer function [1]

$$H(z) = (bb\ z^n + b[n]z^{n-1} + b[n-1]z^{n-2} + \ldots + b[1])/(z^n + a[n]z^{n-1} + a[n-1]z^{n-2} + \ldots + a[1])$$

The assignments produced by the different blocks in Figure 6.6 result in the difference-equation system

input = (*given function of* t)
output = x[n] + bb * input
x[1] = b[1] input − a[1] output)
x[2] = x[1] + b[2] input − a[2] output
. . . .
x[n] = x[n-1] + b[n] input − a[n] output

```
-- SIMPLE ANALOG FILTER
-- H = 1/(s^2 + a[1]s + a[2])
-- successive samples of the filter output are stored
-- in an array that will be used to compute the FFT
-----------------------------------------------------------------
display N14 | display C7 | display R | -- display
NN = 16384 | DT = 0.00001 | TMAX = 2
n = 2 | STATE x[n] | ARRAY a[n]+b[n]+b0[1] = ab
ARRAY OUTPUT[NN], OUTPUTy[NN] | -- arrays for FFT
-----------------------------------------------------------------
-- specify filter parameters
a[1] = 20000 | -- squared circular frequency
a[2] = 40 | -- damping coefficient
b[1] = 1 | -- other a[i], b[i] default to 0
b0[1] = 0 | -- feedforward coefficient
-----------------------------------------------------------------
t = 0 | -- (default initial t would be t = 1)
x[1] = 1 | -- to get impulse response
scale = 0.005 | -- display scale
drunr | -- drunr resets t = 0
write 'type go for FFT' | STOP
-----------------------------------------------------------------
FFT F, NN, OUTPUT, OUTPUTy
scale = 2 | NN = 251
drun SECOND | -- amplitude/phase display
-----------------------------------------------------------------
DYNAMIC
-----------------------------------------------------------------
input=0 | -- for impulse response
```

FIGURE 6.5a Complete program simulating an analog bandpass filter. The main DYNAMIC program segment generates the filter impulse response by setting **input = 0** and the initial value **x[1] = 1**. A **store** operation (Sec. 6.8.1) saves the impulse response in the array **OUTPUT**. The experiment protocol then uses

FFT F, NN, OUTPUT, OUTPUTY

(Sec. 8.4.1) to compute the real part of the complex frequency response in the array **OUTPUT**, and the imaginary part in the array **OUTPUTy**. To display the amplitude and phas response, an extra DYNAMIC segment labeled **SECOND** then uses **get** to retrieve the real and imaginary parts of the frequency response as time histories **xx(t)** and **yy(t)** and computes the amplitude and response. **r = sqrt(xx * xx + yy * yy)** and the phase response **phi = 0.5 * atan2(yy,xx)**. (Continued)

output = x[n] + b0[1] * input | -- note feedforward term
Vectr d/dt x = x{-1} + b * input - a * output
dispt output
---------- fill FFT array
store OUTPUT = output

---- amplitude/phase display
label SECOND
get xx = OUTPUT | get yy = OUTPUTy | -- FFT arrays
r = sqrt(xx * xx+yy * yy)
phi = 0.5 * atan2(yy,xx)
dispt r, phi

FIGURE 6.5a (*Continued*) Complete program simulating an analog bandpass filter. The main DYNAMIC program segment generates the filter impulse response by setting **input = 0** and the initial value **x[1] = 1**. A **store** operation (Sec. 6.8.1) saves the impulse response in the array **OUTPUT**. The experiment protocol then uses

FFT F, NN, OUTPUT, OUTPUTy

(Sec. 8.4.1) to compute the real part of the complex frequency response in the array **OUTPUT**, and the imaginary part in the array **OUTPUTy**. To display the amplitude and phas response, an extra DYNAMIC segment labeled **SECOND** then uses **get** to retrieve the real and imaginary parts of the frequency response as time histories **xx(t)** and **yy(t)** and computes the amplitude and response. **r = sqrt(xx * xx + yy * yy)** and the phase response **phi = 0.5 * atan2(yy,xx)**.

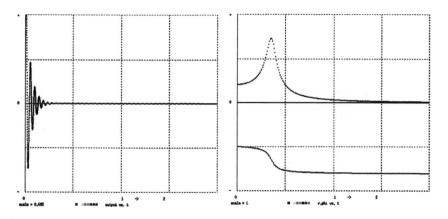

FIGURE 6.5b The impulse response, and the amplitude and phase of the complex frequency response function of the bandpass filter.

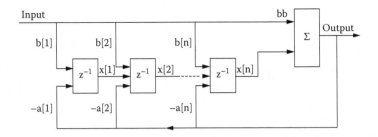

FIGURE 6.6   Block diagram of a digital filter with the z transfer function

$H(z) = (\text{bb } z^n + b[n]z^{n-1} + b[n-1]z^{n-2} + ... + b[1])/\{z^n + a[n]z^{n-1} + a[n-1]z^{n-2} + ... + a[1]\}$

A simulation program repeatedly executes these assignments in order, with **t** successively set to **t = t0, t0+COMINT, t0+2 COMINT,** ... on the right-hand side of each assignment. This updates the assignment targets on the left for the next sampling time. For **t = t0**, the right-hand expressions are initialized with the given initial values of x**[1]**, **x[2]**, ..., **x[n]**, which usually default to 0.

Vectorization again simplifies this program. We declare vectors **x = (x[1], x[2], ..., x[n])**, **a = (a[1], a[2], ..., a[n])**, and **b = (b[1], b[2], ..., b[n])** with

ARRAY x[n], a[n], b[n]

in the experiment protocol and then program

input = (*given function of* **t)**
output = x[n] + bb * input
Vector x = x{-1} + b * input − a * output

in a DYNAMIC program segment. To obtain the filter response to a unit impulse at **t = 0,** we set **t = 0** in the experiment protocol and use **input = swtch(1 - t)**. The filter frequency response is then computed and displayed as in Section 6.4.3.1. The user programs **dcmb-fft*.src, dfilt.src,** and **filter*.src** on the book CD are examples.

6.5 **DOT** PRODUCTS, SUMS, AND VECTOR NORMS

6.5.1 **DOT** Products and Sums of **DOT** Products

Given two vectors **u**, **v** of equal dimension **n**, the DYNAMIC-segment statement

DOT alpha = u * v

assigns the *inner product* of **u** and **v** to the scalar **alpha**, that is,

$$\mathbf{alpha} = \sum_{k=1}^{n} u[k] * v[k]$$

The compiler "unrolls" the summation loop, so that there is, once again, no runtime loop overhead. Nonconformable matrices are rejected with an error message.

As in Section 6.3.2.1, the second vector **v** of a **DOT** product can be a suitably conformable matrix product **A * b** or **A% * b**, as in

DOT beta = y * W * x | DOT beta = x * W% * y

DOT assignments can also produce *sums of inner products* such as

DOT gamma = u1 * v1 + u2 * A * v2 + u3 * v3 + . . .

Vectors and matrices in **DOT** assignment cannot be expressions; they must be defined by earlier assignments.

6.5.2 Euclidean Norms

DOT assignments efficiently compute squared vector norms such as error and energy measures for optimization studies. In particular,

DOT xnormsq = x * x

produces the square **xnormsq** of the *Euclidean norm* ||**x**|| of a vector **x**. The *Euclidean distance* between two vectors **x, y** is the norm ||**x − y**|| of their difference. Thus

Vector error = x - y | DOT enormsq = error * error

results in

$$\mathbf{enormsqr} = \sum_{k=1}^{n} (x[k] - y[k])^2$$

again without summation-loop overhead.

6.5.3 Simple Sums: Taxicab and Hamming Norms

Given a vector **x**, the special **DOT** assignment

DOT xsum = x * 1

efficiently produces *the sum* **xsum = x[1] + x[2] + ...** *of all the elements of* **x**. The notation designates the **DOT** product of **x** and a vector of unity elements, but the multiplications by 1 are not actually performed. As an example,

Vector y = exp(x) | DOT S = y * 1

generates $S = \exp(x[1]) + \exp(x[2]) + \ldots$. Similarly,

Vector y = abs(x) | DOT anorm = y * 1

produces the *taxicab norm* (city-block norm) **anorm = |(x[1])| + |(x[2])| +** **...** of a vector **x**. The taxicab norm of a vector difference (*taxicab distance*, as in a city with rectangular blocks) is another useful error measure.

If all elements of a vector **x** equal 0 or 1 the taxicab norm reduces to the *Hamming norm*, which simply counts all elements equal to 1. The *Hamming distance* **||x - y||** between two such vectors is the count of corresponding element pairs that differ.

6.6 MORE VECTOR/MATRIX OPERATIONS

6.6.1 Vector Difference Equations

The simplest vector difference equations have the form

Vector x = x + f(t;y,z,..., alpha,...)

or equivalently **Vector delta x = f(t; y,z,...,alpha,...)**

which compiles into

x[i] = x[i] + f(t; y[i], z[i], ..., alpha, ...) (i =1,2, ..., n)

More general recursive vector assignments

Vector x = f(t; x, z, ..., alpha, ...)

are legal unless they invoke a matrix-vector product **A * x** or an index-shifted vector **x{k}** with **k > 0**. That returns an error message, for either operation would replace **x[1], x[2]**, ... with new values while the old values are still needed to complete the operation. One can easily circumvent this problem by replacing **x** with a new vector **xx** defined previously with **Vector xx = x.**[*]

6.6.2 Dynamic-Segment Matrix Operations

6.6.2.1 Vector Products

The product of an **m**-dimensional vector **u** and **n**-dimensional vector **v** is the **m**-by-**n** matrix produced by the assignment

MATRIX W = u * v with elements W[i, k] = u[i] * v[k]

This can be used in matrix sums (Section 6.6.2.2). **u**, **v**, and **W** must be declared in the experiment-protocol script; an error is returned if the dimensions do not match.

6.6.2.2 Simple *MATRIX* Assignments

DYNAMIC-segment matrix assignments

MATRIX W = *matrix expression*

do not admit as general expressions as vector assignments do. A few legal examples are

MATRIX W = alpha	**(W[i, k] = alpha)**
MATRIX W = alpha * A	**(W[i, k] = alpha * A[i,k])**
MATRIX W = recip(A)	**(W[i, k] = 1/A[i, k])**
MATRIX W = cos(A)	**(W[i, k] = cos(A[i, k]))**

and also sums of such terms. The Desire Reference Manual has details. **MATRIX** assignments have been used mainly for neural-network simulations [3].

---

[*] The Reference Manual on the book CD deals with the special case of subvectors **x**.

6.6.2.3 A More General Technique

Declaration of *equivalent matrix and vector arrays* with

ARRAY W[m, n] = v

in the experiment protocol (Chapter 5, Section 5.4.2.2) makes it possible to manipulate an **m**-by-**n** matrix as an **mn**-dimensional vector. One can then *assign very general expressions* (Section 6.3.1) *to matrices* and also program *matrix differential equations* and *matrix difference equations*. For example,

Vectr d/dt v = - alpha * v + beta * cos(t)

compiles into the **mn** scalar assignments

d/dt v[i] = - alpha * v[i] + beta * cos(t)
  (i = 1, 2, …, mn)

or

d/dt W[i, k] = - alpha * W[i, k] + beta * cos(t)
  (I = 1, 2, …, m; k = 1, 2, …, n)

This is equivalent to the matrix differential equation

(d/dt) W = - alpha * W + beta * cos(t)

More general expressions can be used on the right-hand side.

6.6.3 Submodels with Vectors and Matrices

Submodels (Chapter 4, Section 4.4.5) can employ vectors and matrices as well as scalar variables and parameters. Arrays used as dummy arguments in experiment-protocol **SUBMODEL** declarations must be previously declared, as in*

```
ARRAY v$[1], vv$[1]
SUBMODEL taxicabber(v$, vv$)
  DOT sum = v$ * 1 | sss = 1/sum
  Vector vv$ = sss * v$
```

---

* This submodel taxicab-normalizes **v$** to form a new vector **vv$** (Section 6.5.3).

end

Dummy arrays such as **v$** and **vv$** are not actually used for computation, so that it is economical to set all their dimensions to 1. Dummy-argument names need not end with a dollar symbol (**$**) but, as in Chapter 4, Section 4.4.5, this is convenient.

Before submodels are invoked in a DYNAMIC program segment, the experiment protocol must declare arrays for subscripted variables, vectors, state vectors, or matrices used as invocation arguments. It is legal to use different array dimensions for multiple invocations of the same submodel. Thus, once the vectors **v1**, **vv1**, **v2**, **vv2** are declared with

ARRAY v1[4], vv1[4], v2[7], vv2[y]

the submodel invocation **taxicabber(v1, vv1)** generates code for

DOT sum = v1 * 1 | sss = 1/sum
Vector vv1 = sss * v1

and **taxicabber(v2, vv2)** generates code for

DOT sum = v2 * 1 | sss = 1/sum
Vector vv2 = sss * v2

6.7 MODEL REPLICATION: A GLIMPSE OF ADVANCED APPLICATIONS

6.7.1 Model Replication

6.7.1.1 Vector Assignments Replicate Models

Assume that a vector **r** and two state vectors **x** and **xdot** are declared with

ARRAY r[n] | STATE x[n]. xdot[n]

and let **u = u(t)** be a scalar variable, and **ww** a constant parameter. Then the vector assignments

Vectr d/dt x = xdot
Vectr d/dt xdot = - ww * x − r * xdot + u

effectively *replicate* the model defined by

```
d/dt x[i] = xdot[i]
d/dt xdot[i] = - ww * x[i] – r[i] * xdot[i] + u
```

n times. Note that the **n** replicated models all have the same input **u** but **n** individual damping coefficients **r[i]**.

Significant applications of model replication literally fill another textbook [3]. Here we can only outline the most useful techniques. A number of interesting user examples are on the book CD.

6.7.1.2 Parameter-Influence Studies

Model replication can greatly simplify parameter-influence studies and save much computing time. This is most evident in large multiparameter studies, but a small example will illustrate our point.

The small simulation program in Chapter 1, Section 1.3.6, used the experiment-protocol loop

```
for i = 1 to 5
  r = 10 * i
  drunr
  display 2
next
```

to call **n = 5** simulation runs exercising the linear-oscillator model

```
d/dt x = xdot
d/dt xdot = - ww * x – r * xdot + u
```

with **n** different values of the damping coefficient **r**. The **display 2** statement keeps the resulting **n** time histories on the same display.

With model replication, the experiment protocol in Figure 6.7 specifies the **n** damping-coefficient values with

```
for I = 1 to n
  r[i] = 10 * i
next
```

and then *produces all n time histories in a single simulation run.* Execution is fast because there is no runtime vector-loop overhead.

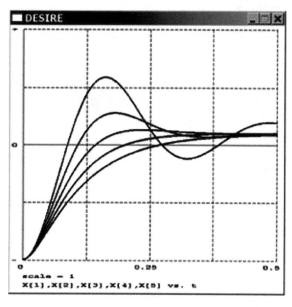

```
--        STEP RESPONSE OF A LINEAR SYSTEM
--            vectorized parameter-influence study
---------------------------------------------------------
display N15 | display C7
TMAX = 0.5 | DT = 0.0001 | NN = 1000
ww = 400 | u = 440 | --              fixed parameters
---------------------------------------------------------
n = 5
ARRAY r[n], X[n] | STATE x[n], xdot[n]
---------------------------------------------------------
for i = 1 to n | --              set parameter values
  r[i] = 10 * i
  next
drun
---------------------------------------------------------
DYNAMIC
---------------------------------------------------------
Vectr d/dt x = xdot
Vectr d/dt xdot = –ww * x–r * xdot + u
-------------------------
Vector X = x – scale | --              offset display
dispt X[1], X[2], X[3], X[4], X[5]
```

FIGURE 6.7 A vectorized parameter-influence study. Each of the **n** models has a different damping coefficient **r**.

6.7.1.3 Vectorized Monte Carlo Simulation

Desire admits up to 40,000 first-order differential equations. One can therefore replicate, say, a 6th-order torpedo model* 5000 times with pseudorandom parameters and/or noise inputs. *One can then take statistics on a sample of 5000 time histories generated by a single simulation run.*

Statistics can be computed by the experiment-protocol script. For example,

DOT sss = error * error | MSI = sss/n

produces the value of a control system mean-square error at the end of a computer run. Many different statistics, including estimates of probabilities and probability densities, can be obtained in this manner [3].

More significantly, a similar DYNAMIC-segment line

DOT sss = error * error | MSI = sss/n

produces *the complete time history of the statistic* **MSI(t)** during the Monte Carlo simulation run.

6.7.1.4 Neural Network Simulation

The output activation (pulse rate) of a simulated neuron with input-synapse activations **x[1], x[2], ..., x[m]** can be modeled with

$$y = \tanh\left(\sum_{k=1}^{m} W[k] * x[k]\right)$$

where **W[1], W[2], ..., W[m]** are the *connection weights* of the **m** input synapses. The soft-limiter function **tanh** is used as a *neuron activation function*. Replication of this model neatly simulates a *neuron layer* as a vector of **n** neuron activations **y[11], y[1], ..., y[n]** fed an *input pattern* represented by the **m**-dimensional vector **x**:

Vector y = tanh(W * x)

---

* User example bigtorpedo.src on the book CD. Under Windows, this Monte Carlo simulation takes 10 minutes on an inexpensive office PC, and half that on a workstation-class PC running Linux.

The output pattern **y** can be the output of another neuron layer. Neuron layers thus readily combine into simulated neural networks, such as the classical three-layer network

Vector v = tanh(W2 * x)
Vector y = W3 * v

Many other types of neuron layers (e.g., softmax layers, radial-basis-function layers) are easily modeled. One *trains* such simulated neural networks by adjusting their connection weights so that a sample of specified network inputs produces desired outputs or output statistics. Training procedures like backpropagation can be implemented with simple matrix difference equations [3].

6.7.2 Other Applications

Model replication has been used to model fuzzy logic control systems [3] and to solve partial differential equations by the Method of Lines [3]. Last but not least, R. Wieland replicated a differential equation model of plant growth model at 1000 or more points of a geographical map. It was then possible to see map regions change from brown to green with the advance of simulated seasons [3].

6.8 TIME HISTORY FUNCTION STORAGE IN ARRAYS

6.8.1 Function Storage and Recovery with **store** and **get**

6.8.1.1 *store* and *get* Operations

The DYNAMIC-segment statements **store** and **get** move time history points into and out of a one-dimensional array, for example, for recovery in subsequent simulation runs, or for Fourier analysis (Section 6.4.3). Specifically,

store X = x

stores successive samples **x(t0), x(t0 + COMINT), x(t0 + 2 COMINT),** ... of a named scalar variable* **x** as corresponding elements **X[1], X[2],** ... of a previously declared vector **X**, with

COMINT = TMAX/(NN −1)

---

* x is here called by name and so cannot be an expression.

```
-- FUNCTION STORAGE AND RETRIEVAL
----------------------------------------------------------------
display N15 | display C7
DT=0.00005 | TMAX=1 | NN=1001
ARRAY X[4 * NN] | -- stores data of 4 runs
w = 5
x = 0.8
drun | drun | drun | drun | -- no reset!
write 'type go to continue' | STOP
--------------
irule 0 | display N14
NN = 4 * NN | t = 0 | TMAX = 4 * TMAX
drun RETRIEVE
----------------------------------------------------------------
DYNAMIC
----------------------------------------------------------------
d/dt x = w * y | d/dt y = - w * x
store X = x
dispt x
----------------------------------------------------------------
   label RETRIEVE
get x = X
dispt x
```

FIGURE 6.8   In this example, **store** saves a harmonic oscillator time history through four continued simulation runs "continued" with **drun**, and a second DYNAMIC program segment uses **get** to display the combined time history of the four runs. Note that the display abscissa correctly ranges between **t** = 0 and **t = 4 TMAX**.

Conversely, the DYNAMIC-segment statement

get x = X

assigns successive elements of a vector **X** to corresponding samples **x(t0), x(t0 + COMINT), x(t0 + 2 COMINT)**, ... of a scalar variable **x. x(t)** is constant between sampling points.

The dimension of the vector **X** normally equals the number **NN** of communication points, or else **store** and **get** simply stop their operation when the either the array or the simulation run is done. **reset** or **drunr** cause the array index to restart at **k = 1**. **get** then reproduces the array

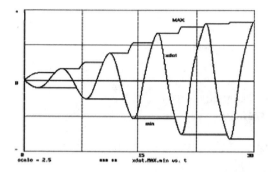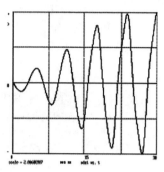

```
--                      AUTOMATIC DISPLAY SCALING
----------------------------------------------------------------
TMAX = 20 | DT = 0.001 | NN = 4500
display N14 | display C7
ARRAY XD[NN] | --        array stores display values
min = 1.0E + 30 | --  initial setting (MAX defaults to 0)
----------------------
x = 0.2  | drunr | --                scaling run, reset
--------------------------------------------
write "type go to continue" |   STOP
scale = MAX | --    second run shows scaled display
drun SCALED
----------------------------------------------------------------
DYNAMIC
----------------------------------------------------------------
d/dt x = xdot | d/dt xdot = –x + (1 – x^2) * xdot
--------
OUT
MAX = abs(xdot) + lim(MAX – abs(xdot))
min = xdot – lim(xdot – min)
store XD = xdot, | --         save xdot for display run
dispt xdot, MAX, min
----------------------------------------------------------------
     label SCALED
get xdot = XD
dispt xdot
```

FIGURE 6.9 Automatic display scaling. The first DYNAMIC program segment solves Van der Pol's differential equation (Chapter 2, Section 2.2.2.1) and tracks the maximum value **MAX** of the absolute value **abs(xdot)**, as described in Chapter 4, Section 4.4.2. The largest observed value of **abs(xdot)** is then used to scale **xdot** in a second DYNAMIC program segment labeled **SCALED**. The program also tracks the minimum value **min** of **xdot**, although it is not needed for scaling.

values from the last simulation run, unless new array values were supplied between runs. If **drun** "continues" a simulation run without **reset**, both **store** and **get** continue properly if the array dimension is large enough. This is demonstrated in the program of Figure 6.8.

6.8.1.2 Application to Automatic Display Scaling
The program of Figure 6.9 combines **store** and **get** operations with maximum tracking (Chapter 4, Section 4.4.2) to produce an *automatically scaled display*. The program employs two DYNAMIC segments (Section 6.4.1). The first DYNAMIC segment solves Van der Pol's differential equation (Chapter 2, Section 2.2.2.1), and the recursive assignment

MAX = abs(xdot) + lim(MAX - abs(xdot))

tracks the maximum excursion **MAX** of |**xdot**|. **store XD = xdot** then saves the **xdot** output for a display run whose display scale **scale** is set to the value of **MAX** at the end of the scaling run.

6.8.2 Time Delay Simulation
The DYNAMIC-segment time delay operations (Figure 6.10)

delay y = DD(x, tau   and   tdelay y = DD(x, tau

are legal only with fixed-step integration and produce delayed output

y(t) = x(t - tau)

by storing and recovering samples of the input **x(t)** in a 1000-element ring-buffer array **DD** declared by the experiment-protocol program with

ARRAY DD[1000]

The nonnegative delay time **tau** can be variable. **delay** samples its input once per communication interval, so that the delay **tau** must be less than **1000 COMINT = 1000 TMAX/(NN-1)**. **tdelay** stores samples at successive **DT** steps. This is preferable with large values of **COMINT** and requires *fixed-step* integration and **tau < 1000 DT**. Delay operations work correctly when a simulation run is "continued" with **drun**.

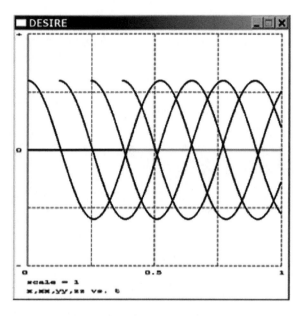

```
--        TIME-DELAY DEMONSTRATION
----------------------------------------------
display N14 | display C7 | display R
DT = 0.0001 | NN = 2000 | TMAX = 1
W = 12
tau1 = 0.125 | tau2 = 0.125 | tau3 = 0.125
x = 0.6
ARRAY XX[1000], YY[1000], ZZ[1000]
drun
----------------------------------------------
DYNAMIC
----------------------------------------------
d/dt x = –w * y | d/dt y = w * x
delay xx = XX(x, tau1
delay yy = YY(xx, tau2
delay zz = ZZ(yy, tau3
dispt x, xx, yy, zz
```

FIGURE 6.10   A simple time delay demonstration.

delay and **tdelay** arguments must be named variables, not expressions. There is no interpolation between samples. One can assign initial values to delayed variables by preloading their delay buffers. Such initial values are not automatically reset by **reset** or **drunr**.

Figures 6.11 and 6.12 show an example application.

```
--  MACKEY-GLASS TIME SERIES
-- try tau=3, 5, 14, 23, and 30
-------------------------------------------------------------------------
irule 1 | --- make sure of fixed-step integration for tdelay!
ARRAY DD[1000] | -- time-delay buffer
x=10 | -- initialize
for i = 1 to 1000 | DD[i] = x | next
-------------------------------------------------------------------------
a = 0.2 | b = 0.1 | c = 10
tau = 14 | -- change delay value for different results!
TMAX = 250 | DT = 0.05 | NN = TMAX/DT + 1
scale = 0.75 | dotscale = 6 | offset = - 0.75 | -- for display
display N14
drun
-------------------------------------------------------------------------
DYNAMIC
-------------------------------------------------------------------------
tdelay Xd = DD(x, tau
xdot = a * Xd/(1 + Xd^c) - b * x
d/dt x = xdot
------------------------------------------------------------- display
X = x + offset | XDOT = dotscale * xdot
dispxy X, XDOT
```

FIGURE 6.11   Program for generating a Mackey-Glass time series.

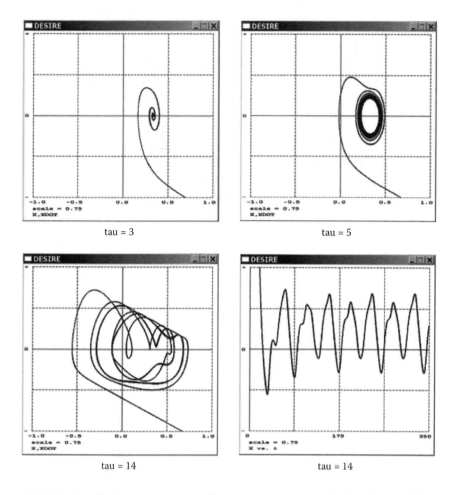

FIGURE 6.12   Phase-plane plots (**xxdot** versus **x**) generated by the Mackey-Glass program for **tau** = 3, 5, and 14, and a time history display.

REFERENCES

1. Dorf, R.C., and R.H. Bishop (2008), *Modern Control Systems*, 11th ed., Pearson Prentice-Hall, Upper Saddle River, NJ.
2. Narendra, K.S., and K. Parthasarathy (1990), Identification and Control of Dynamic Systems Using Neural Networks, *IEEE Trans. Neural Networks*, 1, 4–27.
3. Korn, G.A. (2007), *Advanced Dynamic-System Simulation*, Wiley, Hoboken, NJ.
4. Korn, G.A. (2008), Fast Simulation of Linear Operations: Modeling Analog and Digital Filters with Vectorized State Equations, *Simulation News Europe*, 18/1, April.

CHAPTER 7

Modeling Tricks and Treats

7.1 OVERVIEW, AND A FIRST EXAMPLE

7.1.1 Introduction

Our programs for the eight classical simulation problems described in this chapter introduce the reader to several useful programming tricks. Section 7.1.2 shows how to obtain logarithmic plots by transforming the given differential equations. Sections 7.2.1 through 7.2.3 solve three different problems by calling multiple continuing simulation runs with different parameters. Sections 7.3.1 and 7.3.2 discuss two physiological models, and Sections 7.4.1 and 7.4.2 use multiple DYNAMIC segments for the widely known pilot ejection study. Sections 7.5.1 and 7.5.2 briefly discuss Forrester-type system dynamics and the world-simulation game.

7.1.2 A Benchmark Problem with Logarithmic Plotting

The first of the EUROSIM benchmark problems posed by F. Breitenecker and I. Husinsky [1] models the concentrations **r**, **m**, and **f** of electron-bombarded alkali hydrides with differential equations

$$A = kr * m * f - dr * r \quad B = kf * f * f - dm * m$$
$$dr/d\tau = A \quad dm/d\tau = B - A \quad df/d\tau = p - lf * f - A - 2 * B \quad (7.1)$$

similar to those encountered in population dynamics (Chapter 2, Sections 2.4.1 and 2.4.2) and chemical reaction rate problems. This benchmark problem is not trivial. With the given parameter values, the absolute ratio of the largest to the smallest Jacobian eigenvalue exceeds 120,000 for $\tau =$ **0**, so that we have a very stiff differential equation system, calling for Gear-type integration (Sections A.2.3 and A.2.4 in the Appendix) with small integration steps.

Since the solutions for the given coefficient values vary over a wide range, the benchmark challenge specified a *logarithmic plot* and, for good measure, also a *logarithmic time scale*. Our own program relates the simulation time **t** to the problem time τ so that

$$\tau = 10\,t - t0 \quad d\tau/dt = \ln(10)*(10\char94(10\,t - t0 - 0.00001)) = tt \qquad (7.2)$$

Multiplication of each given differential equation (Equation 7.1) by **tt** then produces new differential equations

$$dr/dt = A * tt \quad dm/dt = (B - A) * tt \quad df/dt = (p - lf * f - A - 2 * B) * tt$$

that neatly solve the problem on a logarithmic time scale. For logarithmic scaling of the state variable **f**, we plot the variable **lgfplus1 = log(e) * ln(f) + 1.**

Serendipitously, our transformation (Equation 7.2) also reduces the problem stiffness (eigenvalue ratio), so that the variable-step Gear integration routine automatically selects larger integration steps. This reclaims some of the extra time needed to compute the exponential factor **tt** repeatedly. Reference 3 has an even faster Desire program that eliminates run-time loop overhead by replacing the seven simulation runs in Figure 7.1 with a single model-replicating run (Section 6.7.1.2).

7.2 MULTIPLE RUNS CAN SPLICE COMPLICATED TIME HISTORIES

7.2.1 Simulation of Hard Impact: The Bouncing Ball

To simulate *abrupt state changes*, it can be useful to terminate a simulation run with a **term** statement (Chapter 2, Section 2.1.3) and then to continue the simulation with a new run using modified initial conditions and/or parameters. Sections 7.2.1 through 7.3.1 exhibit three interesting examples.

-- EUROSIM COMPARISON PROBLEM (LITHIUM CLUSTER)

```
-----------------------------------------------------------------
display N14 | display C7 | display R
irule 16  |  ERMAX = 0.00001 |  --       Gear integration
t0 = 3  |  --                          SHIFT LOG TIME SCALE
TMAX = 1 + t0 | NN = 6000 | DT = 0.0001 | scale = 2
-----------------------------------------------------------------
ln10 = ln(10) |  loge = 1/ln10
kr = 1  |  kf = 0.1  |  dr = 0.1  |  dm = 1  |  --    coefficients
p = 0
f = 9.975 |  m = 1.674 |  r = 84.99 |  --        initial values
-----------------------------------------------------------------
lf = 50 |  drunr |  display 2 |   --    run and reset, keep
display
lf = 100  |  drunr |  lf = 200  |  drunr |  lf = 500  |  drunr
lf = 1000 |  drunr |  lf = 5000 |  drunr |  lf = 10000 | drun
-----------------------------------------------------------------
DYNAMIC
-----------------------------------------------------------------
A = kr * m * f - dr * r | --  these are used several times
                        - precompute!
B = kf * f * f - dm * m
tt = ln10 * (10^(t - t0 - 0.00001)) | --   make log time scale
--
d/dt r = A * tt |   d/dt m = (B - A) * tt | -- state equations
d/dt f = (p - lf * f - A - 2 * B) * tt
--------------------------------------------------- logarithmic plot
lgfplus1 = loge * ln(f) + 1 |   dispt lgfplus1
```

FIGURE 7.1a Complete program for the Lithium-cluster-benchmark simulation.

The simulation of a bouncing ball in Figure 7.2 is a case in point. Instead of attempting to model large semielastic impact forces as such, we terminate the simulation run with

term - y

as soon as the ball falls below **y = 0**. We then start a new simulation run with the current **x** and **y** values, resetting the initial horizontal and vertical velocity components to

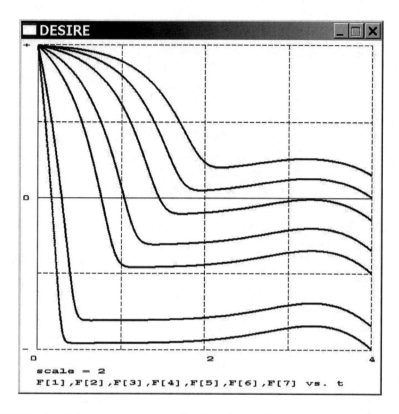

FIGURE 7.1b   Solution time histories for the Lithium-cluster-benchmark problem.

xdot = k * xdot   ydot = - k * ydot

This reproduces the effect of semielastic impact. To keep **term** from stopping the new run immediately, we start **y** at a small positive value instead of 0.

An experiment-protocol **repeat** loop (Chapter 5, Section 5.2.3) repeats this process until **x > 0.6** when a run terminates. A **display 2** statement lets the program plot the continuing runs on the same display. Note that **drunr** does not reset the simulation time **t**, which thus correctly measures spliced-run time.

Figure 7.2 shows the spliced trajectory. It is a good idea to repeat this simulation with a smaller integration step **DT** to make sure that **DT** was small enough to compute the impact time accurately (recall that **term** acts only at the end of integration steps; no **step** statement is needed).

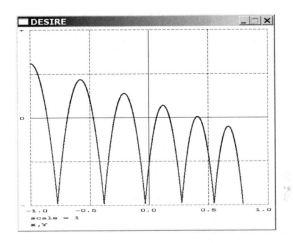

```
--                              BOUNCING BALL SIMULATION
-------------------------------------------------------------------
display N14 | display C7 | --            sets display colors
scale = 1 | DT = 0.002 | TMAX = 1 | NN = 1500
k = 0.94 | --                      restitution coefficient for impact
g = 32.2 |   --                          acceleration of gravity
ÿ = 0.8 | --                       initial conditions (ydot = 0)
x = – 1 | xdot= 1
drun | display 2 | --                    first run, keep display
----------
repeat
   xdot = k * xdot | ydot = – k * ydot | --   rebound velocity
   y = 0.0001 | --                 y > 0 to prevent re-termination!
   drun
   until x > 0.6
-------------------------------------------------------------------
DYNAMIC
-------------------------------------------------------------------
d/dt x =xdot |d/dt y =ydot | d/dt ydot = – g
term -y |--                        terminate run on impact
Y=2 * y – 0.98 |--                        offset display
dispxy x,Y
```

FIGURE 7.2 Bouncing-ball simulation. Each bounce is generated by a new simulation run terminated on impact.

7.2.2 The EUROSIM Peg-and-Pendulum Benchmark [1]

Figure 7.3a shows the geometry of a pendulum whose string hits a peg. When the angle **phi** between pendulum string and vertical becomes more negative than the (negative) peg displacement angle **phip,** the peg effectively shortens the string until the pendulum returns to make **phi** > **phip** again. We use the angle **phi** and its derivative **phidot** as state variables in both modes of the motion. **phi** is a continuous function of the time, but **phidot** changes abruptly when the pendulum strikes or leaves the peg.

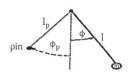

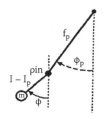

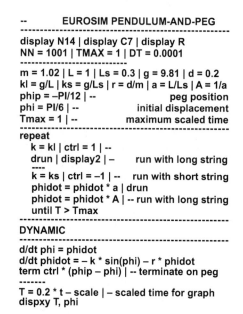

```
--            EUROSIM PENDULUM-AND-PEG
--------------------------------------------------
display N14 | display C7 | display R
NN = 1001 | TMAX = 1 | DT = 0.0001
----------------------------
m = 1.02 | L = 1 | Ls = 0.3 | g = 9.81 | d = 0.2
kl = g/L | ks = g/Ls | r = d/m | a = L/Ls | A = 1/a
phip = -PI/12 | --              peg position
phi = PI/6 | --            initial displacement
Tmax = 1 | --            maximum scaled time
--------------------------------------------------
repeat
   k = kl | ctrl = 1 | --
   drun | display2 | -      run with long string
   ----
   k = ks | ctrl = -1 | --   run with short string
   phidot = phidot * a | drun
   phidot = phidot * A | -- run with long string
   until T > Tmax
--------------------------------------------------
DYNAMIC
--------------------------------------------------
d/dt phi = phidot
d/dt phidot = - k * sin(phi) - r * phidot
term ctrl * (phip - phi) | -- terminate on peg
-------
T = 0.2 * t - scale | - scaled time for graph
dispxy T, phi
```

FIGURE 7.3a   Geometry and program for the pendulum-and-peg simulation.

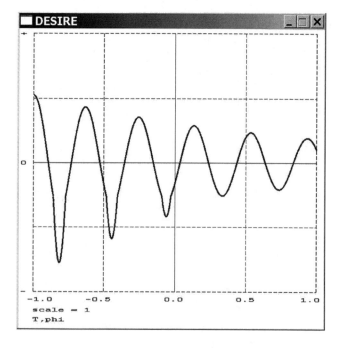

FIGURE 7.3b   Solution time history for the pendulum-and-peg problem generated with successive continued simulation runs.

As in the case of the bouncing-ball impact, the program in Figure 7.3a avoids computation of the impact force exerted by the peg and splices multiple simulation runs. Recalling the pendulum simulation in Chapter 2, Section 2.2.2.2, we model the pendulum state equations with

d/dt phi = phidot
d/dt phidot = - k * sin(phi) - r * phidot

The pendulum starts with a positive initial angle **phi. k = kl = g/L**, where **g** is the acceleration of gravity. We set **ctrl = 1** and terminate the first run when the pendulum hits the peg:

term ctrl * (phip - phi)

Then we start a new run with the shorter string length **Ls**, so that **k = ks = g/Ls**. The damping coefficient **r** is unchanged. The initial angle is **phi = phip** (which is where the last run ended). The new initial angular velocity is

phidot = phidot * L/Ls = phidot * a

which preserves the angular momentum. We now set **ctrl = - 1**, so that the short-string run terminates when the pendulum swings to **phi = phip**. We are then again off the peg and start another long-string run with **k = kl , ctrl = 1**, and

phidot = phidot * Ls/L = phidot * A

This again preserves the angular momentum. An interpreter **repeat** loop continues this process. After some time, damping keeps the pendulum below the peg. The computer time **t** is never reset and again measures spliced-run time. A **display 2** statement lets us plot **phi** versus the scaled time **T = 0.2 * t - scale** over our spliced runs (Figure 7.3b).

7.2.3 The EUROSIM Electronic-Switch Benchmark

The third of Breitenecker's EUROSIM comparison problems [1] simulates the electronic switching circuit (switched amplifier) in Figure 7.4a.

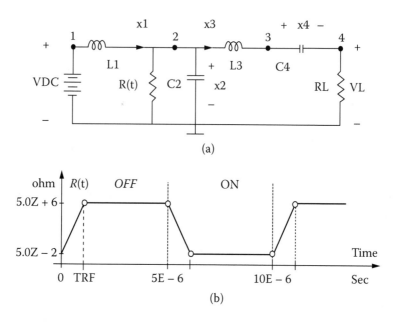

(a)

(b)

FIGURE 7.4   EUROSIM (a) electronic switch, and (b) time history of the switch resistance **R(t)**.

The circuit includes an electronic-switch resistance **R(t)** that changes periodically between **R = 0.05** ohms and **R = 5** megaohms, as shown in Figure 7.4b. Unlike the problems in Sections 7.2.1 and 7.2.2, this simulation involves no truly discontinuous changes, but the resistance **R(t)** in Figure 7.4a changes very rapidly.

Figure 7.5 shows a simulation program and the resulting transient and steady-state behavior of the load voltage **VL** and the current **RI** after **R(t)** starts to switch periodically. We employ the technique of Chapter 4, Section 4.3.1.2, to represent the trapezoidal waveform in Figure 7.5 with the function-generating assignment

R = base + slope * (t - lim(t - TRF) - lim(t - t1) + lim(t - t2))

The program again splices multiple simulation runs, one for each repetition period of **R(t)**. Neither the time nor any state variable is reset between runs. Desire's variable-step integration routines (we use Rule 4 in Table A.1 of the Appendix) automatically terminates the last integration step of each simulation run at the end of the run, and thus at the end of each repetition period.

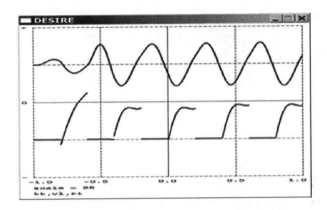

```
                                    EUROSIM SWITCHED AMPLIFIER
--              circuit parameters are scaled so t is in microseconds
-------------------------------------------------------------------
display N14 | display C7 | display R | scale = 25
VDC = 5 | RL = 52.4
L1 = 79.9 | C2 = 17.9E–03 | L3 = 232 | C4 = 9.66E–03
LL1 = 1/L1 | CC2 = 1/C2 | LL3 = 1/L3 | CC4 = 1/C4
-------------------------------------------------------------------
irule 4 | DT = 1.0E–09 | DTMIN = 1.0E–11 | ERMIN = 1.0E–11
NN = 501
-------------------------------------------------------------------
--                              parameters for pulsed resistance
TRF = 1.0E–09
t1 = 5.0 | t2 = t1 + TRF | TMAX = 2*t1 | tmax = 5*t1
base = 5.0E–02 | TRF = 1.0E–09 | slope = (1.0E + 06 – 1.0E–02)/TRF
-------------------------------------------------------------------
t0 = tmax | --                                   offset for xy plot
R = base | --           initial value for track - hold state variable R
drun | display 2 | --                              plot spliced runs
for i = 1 to 4 | t0 = i * t – tmax | t = 0 | drun | next
-------------------------------------------------------------------
DYNAMIC
-------------------------------------------------------------------
VL = x3 * RL | RI = x2/R
d/dt x1 = (VDC – x2) * LL1 | d/dt x2 = (x1 – RI – x3) * CC2
d/dt x3 = (x2 – VL – x4) * LL3 | d/dt x4 = x3 * CC4
step
R = base + slope * (t – lim(t – TRF) – lim(t – t1) + lim(t – t2))
-------------------------------------------------------------------
OUT
vl = VL + 0.5 * scale | ri = 50 * RI–0.5 * scale | --      stripchart plot
tt = t + t0
dispxy tt, vl, ri
```

FIGURE 7.5 Program and solution time history for the EUROSIM electronic switch problem.

7.3 TWO PHYSIOLOGICAL MODELS

7.3.1 Simulation of a Glucose Tolerance Test

Useful physiological-system simulations have studied elaborate models of blood circulation, digestion, and propagation of nutrients and drugs, and control of bodily functions such as blood pressure by drugs

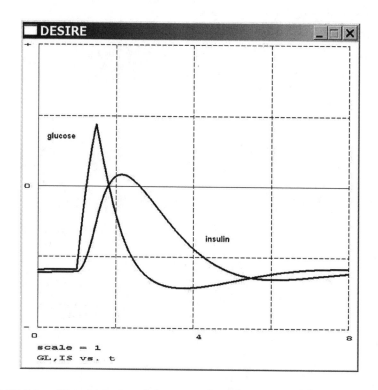

FIGURE 7.6a   Time histories of glucose and insulin masses following a glucose injection.

and fluid intake. We shall discuss a simulation of blood circulation in Section 7.3.2.

The program in Figure 7.6 simulates insulin production in response to a glucose injection under normal and abnormal conditions. The model was designed by a working physician, Dr. J.E. Randall. There are only two state variables, the glucose mass **G** and the insulin mass **I**. The state equation for the glucose mass **G** employs the library function **swtch(t - T1)** from 0 to 1 to turn the glucose injection on, and **swtch(t - T2)** to turn it off. **swtch(G - GK)** switches the "renal spill" of glucose when **G > GK**. The insulin state equation similarly switches insulin production on when **G > G0**.

7.3.2 Simulation of Human Blood Circulation

The PHYSBE model of Figure 7.7, originally developed by J. McLeod [4], is a simplified model of human blood circulation. The welter of different blood vessels is abstracted into seven lumped compartments, namely, two

```
--        GLUCOSE TOLERANCE TEST SIMULATION
--     from Randall, Physiological Simulation
------------------------------------------------------------------------
display N14 |  display C7 |  display R
--
EX = 15000 |  --                extracellular space, ml
CG = 150 |  CI = 150 |  --    glucose and insulin capaci-
tances, ml
--
Q = 8400 |  --             liver glucose release rate, mg/hr
QT = 80000 |  --               glucose infusion rate, mg/hr
--
DD = 24.7 |  --        first-order glucose loss, mg/hr/mg
GG = 13.9 |  --        controlled glucose loss, mg/hr/mg
GK = 250 |  --                   renal threshold, mg
MU = 72 |  --            renal loss rate, mg/hr/mg
G0 = 51 |  --              pancreas threshold, mg
BB = 14.3 |             insulin release rate, mU/hr/mg
AA = 76 |  --        first-order insulin loss, mU/hr/mg
--
G = 81 |  --                 extracellular glucose, mg
I = 5.5 |  --                extracellular insulin, mU
------------------------------------------------------------------------
DT = 1/2000 |  TMAX = 8 |  T1 = 1 |  T2 = 1.5 |  -- time in hrs
NN = 16000
------------------------------------------------------------------------
newvar1 = 0 |  newvar2 = 0 |  -- track-hold state variables
drun
write 'type "go" to continue' |  STOP
reset |  I = 2.9 |  BB = 0.2 * BB |  drun
write "PANCREATIC SENSITIVITY 0.2 TIMES NORMAL"
write 'type "go" to continue' |  STOP
reset |  I = 6.9 |  G = 69.4 |  BB = 10 * BB |  drun
write "PANCREATIC SENSITIVITY 2 TIMES NORMAL"
------------------------------------------------------------------------
DYNAMIC
```

FIGURE 7.6b Program for the glucose-tolerance-test simulation. The user is invited to type go to see different levels of pancreatic sensitivity.   (Continued)

--- **MODEL EQUATIONS**
d/dt G = newvar1 | -- glucose Rate Computation
d/dt I = newvar2 | --- insulin Rate Computation
----
step
newvar1=(Q-G * (GG * I+DD)+QT * (swtch(t-T1)-swtch(t-T2))-swtch(G-GK) * MU *(G-GK))/CG
newvar2 = (-AA * I + swtch(G-G0) * BB * (G-G0))/CI

OUT
GL = 0.005 * G-1 | IS = 0.07 * I-1
dispt GL,IS

FIGURE 7.6b (*Continued*) Program for the glucose-tolerance-test simulation. The user is invited to type go to see different levels of pancreatic sensitivity.

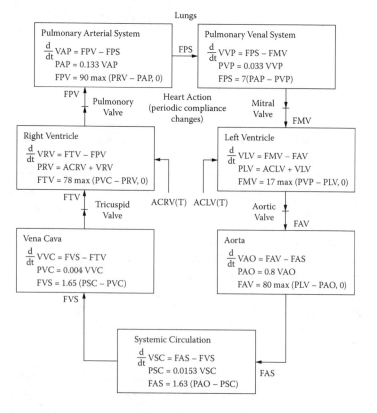

FIGURE 7.7   The PHYSBE blood circulation model.

heart chambers, aorta, vena cava, two lung compartments, and "systemic circulation," which represents body, head, and limbs.

Each compartment has a state equation expressing the effect of inflow and outflow on blood volume,

d/dt bloodVOLUME = FLOWin - FLOWout

and a linearized elasticity equation that relates the compartment blood volume to the compartment blood pressure,

bloodPRESSURE = bloodVOLUME/COMPLIANCE

The flow between adjacent compartments is driven by the pressure difference

**FLOWcompartment_1_to_2 =
ADMITTANCE * (PRESSURE1 - PRESSURE2)**

But where a heart valve separates two compartments there is only *one-way flow*, that is,

**FLOWcompartment_1_to_2
= ADMITTANCE * lim(PRESSURE1 - PRESSURE2)**

as implemented by the library function **lim(x)** (Chapter 4, Section 4.3.1.2). Volumes are measured in milliliters, flow in milliliters per second, and pressure in millimeters of mercury.

Figure 7.8 shows a complete simulation program, and Figure 7.9 presents results. To represent the heart pumping action, the reciprocals **ACRV** and **ACLV** of the two ventricle compliances **CRV** and **CLV** are periodic functions generated by two table-lookup/interpolation function generators (Chapter 4, Section 4.2.2), both driven by a sawtooth time-base generator similar to that described in Chapter 4, Section 4.4.4.1.

PHYSBE-type blood-circulation models can be refined for much greater realism. Large numbers of compartments have been used, and the effects of bleeding, heart failures, blood-pressure-controlling drugs, bypass operations, and weightlessness in a space vehicle have been studied.

```
-- PHYSBE BLOOD CIRCULATION MODEL
--------------------------------------------------------------------------
display N14 | display C7 | display R
--
--                                                    Function Tables
ARRAY FAA[16],FBB[16]
data 0,.0625,.125,.1875,.25,.3125,.375,.4375
data 0.0066,0.16,.32,.45,.625,.78,.27,.0066
--
data 0,.0625,.125,.1875,.25,.3125,.375,.4375
data .0033,.41,.63,.73,.8,.76,.25,.0033
read FAA,FBB
--
x=1 | -- initialize signal generator
--
-- initialize track-hold state variables
p=0
VRV=91 | VAP=220 | VVP=613
VLV=373 | VAO=69 | VSC=2785 | VVC=450
FTV=0 | FPV=0 | FMV=0 | FAV=0
--------------------------------------------------------------------------
DT=0.0001 | TMAX=3 | NN=10001 | scale=100
drun
--------------------------------------------------------------------------
DYNAMIC
--------------------------------------------------------------------------
T=0.5*(p*x+1) | -- sawtooth time
SRV=func1(T;FAA) | SLV=func1(T;FBB)
PRV=VRV*SRV | PAP=0.133*VAP | PPV=0.033*VVP
PLV=VLV*SLV   |    PAO=0.8*VAO    |    PSC=0.0153*VSC   |
PVC=0.004*VVC
FPS=7*(PAP-PPV) | FAS=1.63*(PAO-PSC) |
FVS=1.65*(PSC-PVC)
--
d/dt x=2*p | -- signal generator
d/dt VRV=FTV-FPV | d/dt VAP=FPV-FPS | d/dt VVP=FPS-FMV
d/dt VLV=FMV-FAV | d/dt VAO=FAV-FAS | d/dt VSC=FAS-FVS
d/dt VVC=FVS-FTV
---------------------
 step
 p=sgn(p-x) | -- signal generator
 FTV=78*lim(PVC-PRV) | FPV=90*lim(PRV-PAP)
 FMV=17*lim(PPV-PLV) | FAV=80*lim(PLV-PAO)
--------------------------------------------------------------------------
 pao=PAO-scale | plv=PLV-scale | -- offset runtime display
 dispt pao,plv
```

FIGURE 7.8 Program for the glucose-tolerance-test simulation. The user is invited to type go to see different levels of pancreatic sensitivity

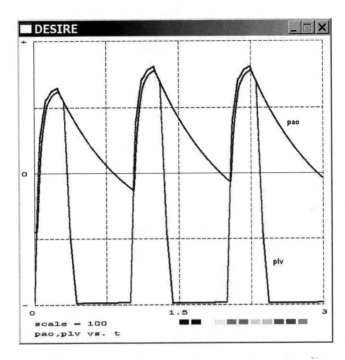

FIGURE 7.9   Solution time histories for the blood pressure in the aorta (**pao**) and the left heart ventricle (**plv**).

7.4 A PROGRAM WITH MULTIPLE DYNAMIC SEGMENTS

7.4.1 Crossplotting Results from Multiple Runs: The Pilot Ejection Problem

The following Desire version of another classical simulation study [5] demonstrates a fairly sophisticated experiment-protocol script using two DYNAMIC program segments. Figure 7.10 shows a pilot and seat being ejected from a fighter plane. The higher air density at lower aircraft altitudes **h** can slow pilot and seat enough to cause a fatal collision with the plane's vertical stabilizer. The study objective is to find the lowest altitude for safe ejection at each aircraft speed **va**.

The experiment protocol in Figure 7.11 calls successive simulation runs that exercise the unlabeled DYNAMIC program segment for different values of **va** and **h**. Each run computes a new ejection trajectory (Figure 7.12a), much as in the cannonball simulation in Chapter 2, Section 2.3.1. Each run terminates when the pilot either clears the aircraft tail or collides with it. Two nested **repeat** loops increment **va** and **h**.

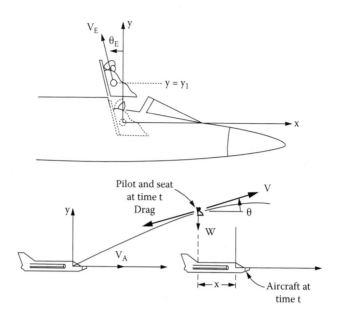

FIGURE 7.10   Geometry for the pilot ejection study.

The outer repeat loop also stores the resulting safe combinations of **va** and **h** in two arrays, **H** and **VA**. **drun crossplt** then exercises a second DYNAMIC program segment labeled **crossplt** to cross-plot the **H[i]** and **VA[i]** values defining the safe-ejection boundary (Figure 7.12b).

crossplt has no derivative assignments, so that **t** takes the values **t = 1, 2, ... , NN = 17**. The variables **Va** and **Ha** thus take successive values **Va(1)**, **Va(2)**, ... and **Ha(1)**, **Ha(2)**... . The **get** statements (Chapter 6, Section 6.8.1.1)

get Va = VA get Ha = H

retrieve successive values **Va(1) = VA[1], Va(2) = VA[2], ...** and **Ha(1) = H[1], Ha(2) = H[2], ...** for crossplotting.

7.5 FORRESTER-TYPE SYSTEM DYNAMICS

7.5.1 A Look at System Dynamics

W. Forrester coined the term *system dynamics* for the application of state-transition models to a variety of business, economic, and social systems, a very significant innovation at the time [6–9]. Addressing an audience

PILOT EJECTION AND CROSSPLOT

```
display N14 | display C7 | display R
DT = 0.002 | TMAX = 2 | NN = 800
---
g = 32.2 | -- gravity acceleration, ft/sec
m = 7 | -- mass of pilot and seat, slugs
y1 = 4 | -- vertical rail height, ft
theta = 15/57.2958 | -- ejection angle, radians
veject = 40 | -- ejection velocity, ft/sec
CD = 1 | -- drag coefficient
h = -50 | -- initial altitude will be 0
va = 100 | -- initial aircraft speed, ft/sec
xtail = 30 | ytail = 20 | -- obstruction (tail) coordinates
rho = 2.377 | -- initial relative air density
C = 0.5E-02 * CD/m | -- drag factor
```

```
-- Table: Relative Air Density RHO versus h
ARRAY RHO[24]
data 0, 1E+3, 2E+3, 4E+3, 6E+3, 1E+4, 1.5E+4, 2E+4, 3E+4,
    4E+4, 5E+4, 6E+4
data 2.377, 2.308, 2.241, 2.117, 1.987, 1.755, 1.497, 1.267
data 0.891, 0.587, 0.364, 0.2238
read RHO
```

```
ARRAY H[20], VA[20] | -- storage arrays for crossplot
```

```
x = - y1 * sin(theta)/cos(theta) | y = y1 | -- initial conditions
vx = va - veject * sin(theta) | vy = veject * cos(theta)
```

```
i = 0 | kk = 0 | -- uble loop through trajectory runs
repeat
write "va = ";va | -- outer loop for different velocities
repeat
h = h + 500 | -- inner loop for different altitudes
drun | yf = y | reset | display 2 | -- run, read y = yf
kk = kk + 1
until yf > ytail | -- pilot must clear the tail!
```

FIGURE 7.11   This Desire program for the pilot ejection study uses two DYNAMIC program segments.   (*Continued*)

```
i = i + 1  |  H[i] = h  |  VA[i] = va  |  -- record safe altitude and
    velocity
va = va + 50  |  vx = vx + 50  |  --
until va > 900
-------
write kk;' runs done; type go for crossplot' | STOP
---------------------------------------------------------------------
display 1      |  -- for crossplot run
TMAX = 1  |  NN = 17  |  drun crossplt  |  -- no derivatives!
---------------------------------------------------------------------
DYNAMIC
---------------------------------------------------------------------
-- unlabeled DYNAMIC segment
rho=func1(y+h,RHO)  |  -- compute air density
CV = C * rho * sqrt(vx^2 + vy^2)  |  -- CV = C * rho * v
d/dt vx = - CV * vx  |  d/dt vy = - CV * vy-g
d/dt x = vx  |  d/dt y = vy
term va * t - x - xtail  |  -- terminate run at or over obstruction
---------------------------------------------------------------------
X = 0.003 * x - 0.95  |  Y = 0.075 * y - 1  |  -- runtime trajectory
    plot
dispxy X, Y
---------------------------------------------------------------------
 label crossplt  |  -- labeled DYNAMIC segment
get Va = VA  |  -- get Va and Ha from the storage arrays VA
    and H
get Ha = H
vv=0.002*Va-scale  |  hh=0.000025*Ha-scale
dispxy vv, hh
```

FIGURE 7.11 (*Continued*) This Desire program for the pilot-ejection study uses two DYNAMIC program segments.

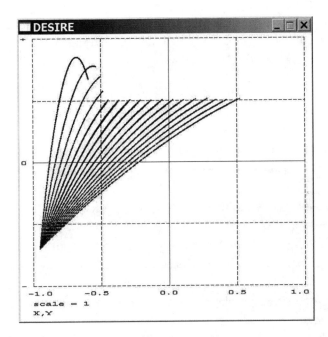

FIGURE 7.12a   Pilot ejection trajectories. Note how the unsafe trajectories are stopped by collision with the aircraft elevator.

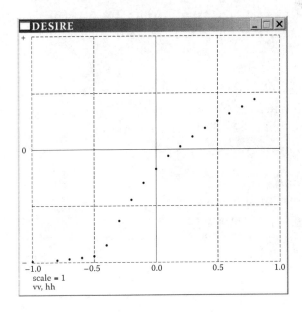

FIGURE 7.12b   Crossplot of safe ejection altitudes vv for different aircraft speeds **vv.**

of business managers, Forrester used state difference equations rather than differential equations, and constant time increments **Δt** (e.g., a day, a month, or a year), so that

$$xi(t + Δt) = xi(t) + fi(t, x1, x2, ..., xn) (i = 1, 2, ..., n)$$

with defined variables and system parameters introduced as in Chapter 1, Section 1.1.2. Variables and parameters were often integers, but they were represented by continuous variables, just as in Sections 1.3.4 and 1.3.5 (Chapter 1).

System-dynamics simulations have been successfully applied to a wide variety of problems [6–8], and at least two special simulation languages (notably Stella) were written specially for this purpose. It is fair to say, though, that one can easily program system-dynamics simulations with any ordinary continuous-system simulation language. In Desire notation, one simply replaces the difference equations with derivative assignments

$$d/dt \ xi = fi(t, x1, x2, ..., xn) (i = 1, 2, ..., n)$$

and uses Euler integration (**irule 2**) with **DT = Δt**. Actually trying different integration rules might produce interesting new models.

7.5.2 World Simulation

Most of Forrester's simulations dealt with down-to-earth business problems [6–8], but Reference 9 presents a provocative model that depicts the past, present, and future of the entire world in terms of only 5 (!) state variables, namely,

Population, **P**
Amount of natural resources, **NR**
Amount of capital investment, **CI**
Amount of pollution, **POL**
Fraction of capital investment devoted to agriculture, **CIAF**

Figure 7.13 shows typical results, and Figure 7.14 shows a complete Desire program for Forrester's world model, and Figure 7.15 shows typical results. Space limitations force us to refer you to the long list of symbol definitions in Reference 9 (our program uses the same symbols as the

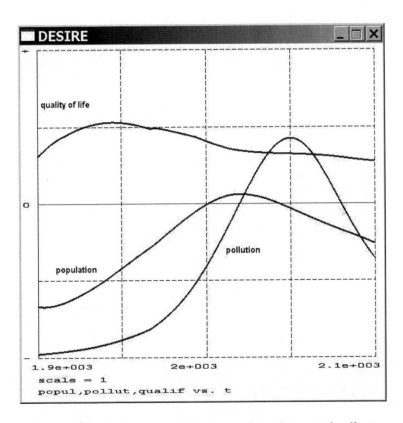

FIGURE 7.13   World simulation time histories of population and pollution, and of the authors' "quality of life" function from 1900 to 2100.

reference). The model is, in essence, a population-dynamics model of the type discussed in Chapter 2, Section 2.4.2, with the birth and death rates determined by a complicated (and controversial) set of functions of the resource, investment, and pollution variables. Many different features can be added at will.

Desire experiment-protocol scripts make it easy to enter table values for the 22 different functions with **data/read** assignments (Chapter 5, Section 5.3.2.2). Figures 7.14 and 7.15 are based on the exact function values given in Reference 9. Modern personal computers are so fast that interactive experiments with different investment and death rates will soon make you feel like a microcosmic god. Such models can be tuned (and have been tuned) to support any and all political viewpoints.

--WORLD II GAME (based on Forrester, World Dynamics)
--
-- "ERITIS SICUT DEUS, SCIENTES BONUM ET MALUM"

-- Function Tables

ARRAY BRMM[12],NREM[10],DRMM[16],DRPM[14],
DRFM[16],DRCM[12]
ARRAY BRCM[12],BRFM[10],BRPM[14],FCM[12],
FPCI[14],CIM[12]
ARRAY FPM[14],POLCM[12],POLAT[14],CFIFR[10],QLM[12
],QLC[22]
ARRAY QLF[10],QLP[14],NRMM[22],CIQR[10]
--
data 0,1,2,3,4,5;1.2,1,.85,.75,.7,.7 | -- BRMM
data 0,.25,.5,.75,1;0,.15,.5,.85,1 | -- NREM
data 0,.5,1,1.5,2,2.5,3,3.5;3,1.8,1,.8,.7,.6,.53,.5 | -- DRMM
data 0,10,20,30,40,50,60;.92,1.3,2,3.2,4.8,6.8,9.2 | -- DRPM
data 0,.25,.5,.75,1,1.25,1.5,1.75;30,3,2,1.4,1,.7,.6,.5 | -- DRF
data 0,1,2,3,4,5;.9,1,1.2,1.5,1.9,3 | -- DRCM
data 0,1,2,3,4,5;1.05,1,.9,.7,.6,.55 | -- BRCM
data 0,1,2,3,4;0,1,1.6,1.9,2 | -- BRFM
data 0,10,20,30,40,50,60;1.02,.9,.7,.4,.25,.15,.1 | -- BRPM
data 0,1,2,3,4,5;2.4,1,.6,.4,.3,.2 | -- FCM
data 0,1,2,3,4,5,6;.5,1,1.4,1.7,1.9,2.05,2.2 | -- FPCI
data 0,1,2,3,4,5;.1,1,1.8,2.4,2.8,3 | -- CIM
data 0,10,20,30,40,50,60;1.02,.9,.65,.35,.2,.1,.05 | -- FPM
data 0,1,2,3,4,5;.05,1,3,5.4,7.4,8 | -- POLCM
data 0,10,20,30,40,50,60;.6,2.5,5,8,11.5,15.5,20 | -- POLAT
data 0,.5,1,1.5,2;1,.6,.3,.15,.1 | -- CFIFR
data 0,1,2,3,4,5;.2,1,1.7,2.3,2.7,2.9 | -- QLM
data 0,.5,1,1.5,2,2.5,3,3.5,4,4.5,5;2,1.3,1,.75,.55,.45,.38
data .3,.25,.22,.2 | -- QLC
data 0,1,2,3,4;0,1,1.8,2.4,2.7 | -- QLF
data 0,10,20,30,40,50,60;1.04,.85,.8,.3,.15,.05,.02 | -- QLP
data 0,1,2,3,4,5,6,7,8,9,10;0,1,1.8,2.4,2.9,3.3,3.6,3.8,3.9
data 3.95,4 | -- NRMM
data 0,.5,1,1.5,2;.7,.8,1,1.5,2 | -- CIQR
--

FIGURE 7.14   An experiment-protocol script for the world model. Parameter values are based on Reference 9, which discusses this model in detail.

(Continued)

```
read BRMM,NREM,DRMM,DRPM,DRFM,DRCM,
BRCM,BRFM,BRPM,FCM,FPCI,CIM
read FPM,POLCM,POLAT,CFIFR,QLM,QLC,QLF,QLP,
NRMM,CIQR
--
-------------------------------------------------------------
-- Constant Parameters
-------------------------------------------------------------
BRN = 0.04
DRN = 0.028
LA = 1.35E+08
PDN = 26.5
FC1 = 1
FN = 1
CIGN = 0.05
CIDN = .025
CIAFT = 15
CIAFN = 0.3 | -------------- = 3 IN ORIGINAL
NRUN = 1
POLS = 3.6E+09
POLN = 1
QLS = 1
NR1 = 9.0E+11
-------------------------------------------------------------
-- Initial Values
-------------------------------------------------------------
P = 1.65E+09
NR = 9.0E+11
CI = 0.4E+09
POL = 0.2E+09
CIAF = 0.2
-------------------------------------------------------------
-- make one run
DT = 1
TMAX = 200
t = 1900
NN = 201 | display N15 | display C7 | display R
irule 2 | -- Euler integration
drun
```

FIGURE 7.14 (*Continued*) An experiment-protocol script for the world model. Parameter values are based on Reference 9, which discusses this model in detail.

DYNAMIC

NRFR = NR/NR1 | NRM = func1(NRFR;NREM)

--

CIR = CI/P | CIRA = CIR * CIAF/CIAFN
PO = func1(CIR;POLCM) | FP = func1(CIRA;FPCI)

--

MSL = CIR * NRM * (1 - CIAF)/(1 - CIAFN)
CM = func1(MSL;CIM) | BRM = func1(MSL;BRMM)
DRM = func1(MSL;DRMM)
NM = func1(MSL;NRMM) | QM = func1(MSL;QLM)

--

CR = P/(LA * PDN)
BRC = func1(CR;BRCM) | DRC = func1(CR;DRCM)
FC = func1(CR;FCM) | QC = func1(CR;QLC)

--

POLR = POL/POLS
BRP = func1(POLR;BRPM) | DRP = func1(POLR;DRPM)
FM = func1(POLR;FPM)
PT = func1(POLR;POLAT) | QP = func1(POLR;QLP)

--

FR = FC1 * FC * FM * FP
BRF = func1(FR;BRFM) | CC = func1(FR;CFIFR)
DRF = func1(FR;DRFM) | QF = func1(FR;QLF)

--

QL = QC * QF * QM * QP * QLS | QX = QM/QF
CQ = func1(QX;CIQR)

--

---

d/dt P=P * (BRN * BRF * BRM * BRC * BRP -DRN * DRF *
DRM * DRC * DRP)
d/dt NR = - P * NRUN * NM
d/dt CI = P * CM * CIGN-CI * CIDN
d/dt POL = P * POLN * PO - POL/PT
d/dt CIAF = (CC * CQ - CIAF)/CIAFT

FIGURE 7.15   DYNAMIC program segment for the world model. Functions and parameters are based on Reference 9. (*Continued*)

```
OUT | --                        runtime display
popul = P * 2.0E-10 - scale
pollut = 0.25 * POLR - scale
qualif = 0.5 * QL
dispt popul,pollut,qualif
```

FIGURE 7.15 (*Continued*) DYNAMIC program segment for the world model. Functions and parameters are based on Reference 9.

REFERENCES

1. Breitenecker, F., and I. Husinsky (1995), Results of the EUROSIM Comparison "Lithium Cluster Dynamics," *Proc. 1995 EUROSIM*, Vienna, Austria.
2. Upadhayay, S.K. (2000), *Chemical Kinetics and Reaction Dynamics*, Springer, New York.
3. Breitenecler, F. (1999), Comparisons of simulation tools and simulation techniques, *Simulation News Europe*, 26, July.
4. McLeod, J. (1966), PHYSBE: A Physiological Benchmark Experiment, *Simulation*, 324–329.
5. Simulation Councils Software Committee (1957): The SCI Continuous-system Simulation Language, *Simulation*, December.
6. Forrester, J.W. (1968), *Principles of Systems*, Wright-Allen Press, Cambridge, MA.
7. Forrester, J.W (1961), *Industrial Dynamics*, MIT Press, Cambridge, MA.
8. Forrester, J.W. (1969), *Urban Dynamics*, MIT Press, Cambridge, MA.
9. Forrester, J.W. (1972), *World Dynamics*, MIT Press, Cambridge, MA.

General-Purpose Mathematics

8.1 INTRODUCTION

8.1.1 Overview

Desire is not a substitute for full-featured mathematics programs such as Scilab, or for comprehensive statistics packages such as R. But you can quickly do a good deal of scientific and engineering computation without leaving your simulation desktop. Typed commands like **write 1.7* cos(0.33)** make Desire a very powerful *scientific calculator*. Desire's scripting language can

- Compute expressions and functions (Chapter 2, Section 2.1.1)

- Handle files and device input/output (Chapter 5, Sections 5.4.1 and 5.4.2).

- Produce sums and products of vectors and matrices (Chapter 6, Sections 6.2.3 and 6.2.4)

- Invert matrices and solve linear equations (Chapter 6, Section 6.2.5)

- Compute fast Fourier transforms and convolutions (Sections 8.4.1–8.4.3)

Later in this chapter, we show that our interpreted scripts can also manipulate and plot complex numbers (Sections 8.3.1–8.3.3).

Finally, fast runtime-compiled DYNAMIC-segment code can efficiently evaluate and display a wide variety of functions and statistics (Sections 8.2.1–8.2.5).

8.2 COMPILED PROGRAMS NEED NOT BE SIMULATION PROGRAMS

8.2.1 Dummy Integration Simply Repeats DYNAMIC-Segment Code

We saw in Chapter 3, Section 3.2.2, that Desire automatically defaults to "dummy integration" (**irule 0**) when a DYNAMIC program segment contains no derivative assignments (no **d/dt** or **Vectr d/dt**). **drun** then simply compiles the DYNAMIC-segment code and executes it **NN** times. The system variable **t** takes the **NN** successive values

$$t = t0, t0 + COMINT, t0 + 2 * COMINT, \ldots ,$$
$$t0 + (NN - 1) * COMINT = TMAX$$
$$\text{with} \quad COMINT = TMAX/(NN - 1)$$

If **t0** and **TMAX** are not specified, they default to **t0 = 1** and **TMAX = NN - 1**, so that **t** simply takes the values **t = 1, 2, … , NN**. This conveniently programmed "automatic loop" has many useful applications.

8.2.2 Fast Graph Plotting

8.2.2.1 A Simple Function Plot

The automatic-loop technique quickly generates nicely labeled color graphs. Here is a complete program for plotting the sine and cosine functions (Figure 8.1):

```
NN=251
t = - PI | TMAX = 2 * PI
-------------------------------------
DYNAMIC
-------------------------------------
x = cos(t) | y = sin(t)
dispt x,y
```

We programmed **t = - PI** and **TMAX = 2 * PI** to obtain the minimum and maximum argument values **t = t0 = - π** and **t = t0 + TMAX = π**. It

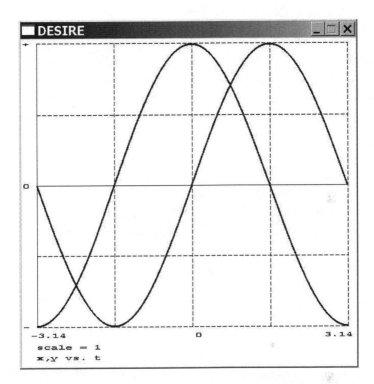

FIGURE 8.1 Full-screen plots of **x = cos(t)** and **y = sin(t)** generated by the program of Section 8.2.2.1.

is easy to change the display scale (e.g., **scale = 2**) and the display colors, or to obtain stripchart-type displays (Chapter 1, Section 1.3.7). To plot **y** against **x**, substitute **dispxy** for **dispt**. One can similarly produce console listings and printouts.

Such programs execute immediately like interpreter programs, but the runtime-compiled code is much faster.

8.2.2.2 Array-Value Plotting

To plot typed, computed, or file-input array values, we use the **get** operation (Chapter 6, Section 6.8.1.2), as in the following example:

```
NN = 12 | ARRAY X[NN], Y[NN] | --    declare two arrays
input X | --       input NN array values X[k] from console
--
--                            read NN Y[k] values from file
connect 'Yfile.dat' as input 3
input #3, Y
```

```
disconnect 3
--
drun
------------------------------------------------------------------------
DYNAMIC
------------------------------------------------------------------------
get x = X  |  get y = Y
dispt x, y
```

Since there are no differential equations, and **TMAX** is not specified, **t** takes successive values **t = 1, 2, ...** to represent the array index. Compiled plotting code is so fast and convenient that it pays to use it even when the rest of a computation is all interpreted script code.

8.2.3 Fast Array Manipulation

Compiled DYNAMIC-segment code manipulates one-dimensional arrays more efficiently than an interpreted script. As an example, the short program

```
NN = 51
ARRAY SECANT[NN]
t = 0  |  TMAX = PI/4
drun
------------------------------------------------------------------
DYNAMIC
------------------------------------------------------------------
secant = 1/cos(t)  |  store SECANT = secant
```

stores **NN = 51** successive samples of the function **secant(t)** for the **NN** uniformly spaced argument values

$$t = 0, \pi/200, 2\pi/200, \ldots, 50\pi/200 = \pi/4$$

Such computations combine nicely with compiled graph plotting (Section 8.2.4) and with compiled or interpreted matrix operations (Chapter 6, Sections 6.2.1–6.2.5). As an example, the small program

```
n = 4  |  ARRAY x[n], y[n]
for k = 1 to n  |  x[k] = k | next
NN = 2
```

drun

DYNAMIC

Vector y = x * x * x
DOT sum = y * 1
type sum

sums successive cubes of **x = 1, 2, . . . , n** twice.[*]

8.2.4 Statistical Computations

Desire is specifically designed for Monte Carlo simulation of dynamical systems and permits efficient computation of statistics used for estimation or statistical tests. This is a large subject, described in a separate textbook [1] that also introduces neural network techniques for statistical regression and pattern recognition.

The most commonly used statistics, including estimates of probabilities and probability densities, are functions of *time averages* and *sample averages*. DYNAMIC program segments can efficiently compute

1. Averages over **t** as differential equation state variables with

$$\textbf{d/dt xxx = x } \mid \textbf{ xavg = xxx/t (t > 0)}$$

using the initial values **xxx(0) = 0, xavg(0) = x(0);** or with

$$\textbf{d/dt xavg = (x - xavg)/t}$$

using the initial values **t = t0 = 1.0E-275, xavg(t0) = x(t0).**

2. Averages over **t** as difference equation state variables (Section 3.3.1) with

$$\textbf{xsum = xsum + x } \mid \textbf{ xavg = xsum/t (t = 1, 2, ...)}$$

---

[*] **NN = 1** is illegal, since it would imply **COMINT = TMAX/(NN - 1) = ∞**. Desire would automatically replace **NN = 1** with the default setting **NN = 251**.

or with

$$xavg = xavg + (x - xavg)/t \ (t = 1, 2, \ldots)$$

using the initial values **xsum(1) = xavg(1) = x(1)**.

3. Averages over arrays (vectors) **(x[1], x[2], ..., x[n])** representing samples of measurements. We use the **DOT** operations described in Sections 6.2.3 and 6.5.1:

$$\textbf{DOT xsum = x * 1 | xavg = xsum/n}$$
$$\textbf{DOT xxsum = x * x | xxavg = xxsum/n}$$

8.3 INTEGERS, COMPLEX NUMBERS, AND INTERPRETER GRAPHICS

8.3.1 **INTEGER** and **COMPLEX** Quantities

Desire treats all undeclared quantities as double-precision (64-bit) real numbers. Experiment-protocol scripts can declare **INTEGER** and **COMPLEX** variables and arrays with up to 10 dimensions:

INTEGER a, b, c , dd[2, 3]
COMPLEX Z[4,4], q[100], z, w

Mixed expressions are legal, and data conversion is automatic. Real expressions assigned to **INTEGER** quantities are rounded to fixed-point format. **INTEGER** function arguments are converted to real numbers. The truncation function **trnc(x)** and the rounding function **round(x)** produce truncated and rounded real numbers, not integers.

input, data/read, and **write** statements, and appropriate user-defined interpreter functions and procedures (Chapter 4, Section 4.2.3; Chapter 5, Section 5.2.4) all admit and recognize **INTEGER** and **COMPLEX** quantities. Data types in function or procedure definitions and calls must match. DYNAMIC program segments, and also interpreted **Vector, MATRIX**, and **DOT** operations (Chapter 6, Sections 6.2.1–6.2.5), admit only real quantities.

8.3.2 Octal and Hexadecimal Integer Conversions

Desire experiment-protocol scripts recognize literal *octal and hexadecimal integers* by the respective *prefixes* **&** and **&&**, as in

Noctal = &17370   **Mhex = &&FC2**

Scripts can use such literal integers in expressions. Currently, octal and hexadecimal integers can be as large as 65535 (decimal) and may encode negative numbers. Desire rejects unacceptable codes and disregards the leading digits if you enter too many digits.

Conversely,

write %12918 and write %%12918

display the decimal integer 12918 in octal and hexadecimal form, respectively.

8.3.3 Complex Number Operations and Library Functions

COMPLEX quantities must be declared before they are used. The first **COMPLEX** declaration automatically defines the imaginary number **j** (square root of −1), so that **COMPLEX** numbers can be assigned values **a + j * b**. The symbol **j** is available for other purposes only before the first **COMPLEX** declaration. One can also enter **COMPLEX** quantities in the form **[a, b]**. Here are some examples:

```
COMPLEX z, c, ww, Z[2,3]
. . . . . . . . . . .
ff = 2.11 | c = 1000
ww = 23.444 - j * ff
c = c + 1 + j * 2 | z = ff * c - ff +[2, ff - 1] - Z[1,2]
```

Note that **c = 1000** in this example is equivalent to **c = 1000 + j * 0**.

Library functions other than **exp(x)**, and also exponentiation (**x^y**), **input**, and **read** operations do *not* work for **COMPLEX x** but return an error message.

The real functions

Re(z)   **Im(z)**   **Cabs(z)**   **Arg(z)**

(real part, imaginary part, absolute value, and argument of z) and the **COMPLEX** library functions

Conjug(z)   **exp(z)**

(*complex conjugate* and *complex exponential* of z) accept **COMPLEX** arguments. The **COMPLEX** function

expj(x) ≡ exp(j * x) ≡ cos(x) + j * sin(x)

is normally used for real arguments **x** only. If **z** is **COMPLEX, expj(z)** returns **exp(j * Re(z))**.

8.3.4   Interpreter Graphics, Complex Number Plots, and Conformal Mapping

8.3.4.1 Interpreter Plots

Interpreted scripts can do *point-by-point plotting*. To this end, one must first set coordinate-net colors with **display C n** and then open the Graph Window with **display A**. The system variable **scale** (which defaults to 1) scales both coordinate axes, as in Chapter 1, Section 1.3.7. Then

plot x, y, cc

plots a point **(x, y)**; **cc** is a color code between 0 and 15. Use **display R** to draw thicker points, and **display F** to clear the screen. As an example,

```
display C7 | display R
display A
for k = 1 to 100 | plot k/100, sin(k/10), 14 | next
```

can be programmed or entered in command mode. Interpreter plots permit quick experiments locating individual points.

8.3.4.2 Complex Number Plots

Interpreter plots can plot complex variables. Figure 8.2 is an example of a *frequency-response plot* that plots the absolute value **Cabs(z)** and the argument (phase angle) **Arg(z)** of a complex impedance **z** against the logarithm of the circular frequency **w**.

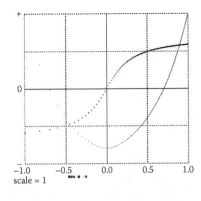

```
--    COMPLEX REQUENCY RESPONSE
--         Z = R + j(wL - 1/wC)
------------------------------------------
display R | display C7
a = 3 | --                         log scale
b = 0.0002 | --      amplitude display scale
c = 0.4 | --              phase display scale
R = 1000 | --                          Ohms
L = 1 | --                           Henries
C = 1.0E-06 | --                       Farads
------------------------------------------
COMPLEX Z
display A
for w = 100 to 10000 step 50
    Z = R + j * (w * L - 1/(w * c))
    plot log(w) - a, b * Cabs(Z) - 0.999, 14
    plot log(w) - a, c * Arg(Z), 12
next
```

FIGURE 8.2 This interpreter loop plots the amplitude **Cabs(z)** and the argument (phase angle) **Arg(z)** of a complex impedance **z = R + j(w L - 1/wC)** against the logarithm of the circular frequency **w**.

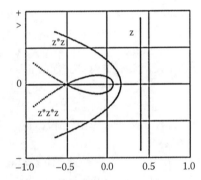

```
--    CONFORMAL MAPPING
------------------------------------------
display C7 | display R
display A
--
COMPLEX z
for y = -0.9 to 0.9 step 0.001
    x = 0.4
    z = x + j * y
    plot z, 10
    plot z * z, 11
    plot z * z * z, 14
next
```

FIGURE 8.3 Conformal mapping of the curve **z = 0.4 + j y** by the functions **z²** and **z³** as **y** changes between -0.9 and 0.9.

After opening the Graph Window with **display C n** and **display A**,

plot z, cc

plots points **z = x + j * y** *in the complex number plane.* As before, **cc** is a color code, and both **x** and **y** axes are scaled by the system variable **scale**. Important applications include *complex frequency-response plots* (Nyquist plots) and *root-locus plots* in control engineering.

As another example, the interpreter program in Figure 8.3 demonstrates *conformal mapping.*

8.4 FAST FOURIER TRANSFORMS AND CONVOLUTIONS

8.4.1 Fast Fourier Transforms

8.4.1.1 Simple Fourier Transformations

Desire Fourier transformations do not manipulate **COMPLEX** quantities. Instead, the programmed or command-mode script statements

$$\textbf{FFT F, n, x, y} \quad \text{and} \quad \textbf{FFT I, n, x, y} \qquad (8.1)$$

operate on two previously declared one-dimensional arrays **x** and **y** of real numbers that represent the real and imaginary parts of complex numbers **x[i] + j y[i]** (i = 1, 2,...). The arrays **x** and **y** must have dimensions at least equal to **n**, which must equal 32, 64, 128, 256, 512, 1024, 2048, 4096, 8192, or 16,384 (these are powers of 2).*

The operations (Equation 8.1) implement in-place *forward or inverse discrete Fourier transformations* of an array of complex numbers **x[i] + jy[i]**, as detailed in Table 8.1. The original arrays **x** and **y** are overwritten.

The script in Figure 8.4 first generates the time-history array **x** for a rectangular pulse; all **y[k]** equal 0 by default. **FFT F, NN, x, y** then replace the **NN** original **x** and **y** values with the real and imaginary parts **x** and **y** of the complex Fourier transform.

drunr and **drun SECOND** next call two compiled DYNAMIC program segments that use **get** operations to produce time histories **xx(t) = x[t]** and **yy(t) = y[t]** (t = 1, 2, ..., **NN**) for display, much as in Section 8.2.2.2. The top graph in Figure 8.4 represents the real part **x** and the imaginary part **y** of the transform. The bottom graph displays the absolute value **r = sqrt(xx^2 + yy^2)** and the argument (phase angle) **phi = atan2(yy, xx)**.

8.4.2 Simultaneous Transformation of Two Real Arrays

The programmed or command-mode statement

FFT R, n, x, y, u, v

simultaneously transforms *two* real input arrays **x** and **y** into *four* real output arrays **x, y, u, v**, all previously dimensioned to a size at least equal to **n**, which must again equal a power of 2 as in Section 8.4.1. The complex

---

* If you have, say, only 900 data points, use 1024-point arrays; they will be automatically padded with 0s. If one of the array sizes exceeds **n**, then the array is effectively truncated to size **n**.

TABLE 8.1   Desire Fast Fourier Transforms

1. **FFT F, NN, x, y** implements the *discrete Fourier transform*

$$x[i] + j\, y[i] \leftarrow \sum_{k=1}^{NN} (x[k] + j\, y[k])\, exp(-2\pi j\, ik/NN)\ (i = 1, 2, \dots, NN)$$

   FFT I, NN, x, y implements the *discrete inverse Fourier transform*

$$x[k] + j\, y[k] \leftarrow (1/NN) \sum_{i=1}^{NN} (x[i] + j\, y[i])\, exp(2\pi j\, ik/NN)\ (k = 1, 2, \dots, NN)$$

2. When the **x[k], y[k]** represent **NN** *time-history samples* taken at the sampling times **t = 0, COMINT, 2 COMINT, . . . , TMAX** with **COMINT = TMAX/(NN - 1)** then the time-domain period associated with the discrete Fourier transform equals **T = NN * COMINT = NN * TMAX/(NN - 1)** (*not* **TMAX**). Approximate frequency-domain sample values of the ordinary *integral* Fourier transform are represented by **x[i] * T/NN, y[i] * T/NN**. Windowing may be needed and can be implemented in either the time or the frequency domain.

3. If the **x[i], y[i]** represent **NN** *frequency-domain samples* taken at the sample frequencies **f = 0, COMINT, 2 COMINT, . . . , TMAX** with **COMINT = TMAX/(NN - 1)** then

t	represents **f** (frequency)
COMINT	represents **1/T** (frequency-domain sampling interval)
TMAX	represents **(NN - 1)/T**
NN * TMAX/(NN-1)	represents **NN/T** (frequency-domain "repetition period")

arrays defined by **x + ju** and **y + jv** are the discrete Fourier transforms of the real input arrays **x** and **y**, which are overwritten. Note again, that we do *not* use the **COMPLEX** data type.

8.4.3 Cyclical Convolutions

The programmed or command-mode statement

FFT C, n, x, y

also works with two real arrays **x, y** dimensioned as before. This operation first Fourier-transforms **x** and **y** and then computes the inverse transform of their product. The real and imaginary parts of this transform overwrite the original arrays **x** and **y**.

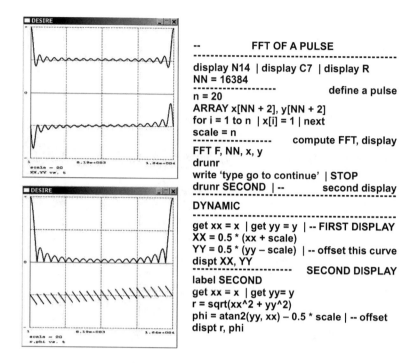

```
--              FFT OF A PULSE
-------------------------------------------------
display N14 | display C7 | display R
NN = 16384
---------------------          define a pulse
n = 20
ARRAY x[NN + 2], y[NN + 2]
for i = 1 to n  | x[i] = 1 | next
scale = n
-------------------          compute FFT, display
FFT F, NN, x, y
drunr
write 'type go to continue' | STOP
drunr SECOND  | --          second display
-------------------------------------------------
DYNAMIC
-------------------------------------------------
get xx = x | get yy = y  | -- FIRST DISPLAY
XX = 0.5 * (xx + scale)
YY = 0.5 * (yy – scale)  | -- offset this curve
dispt XX, YY
------------------------          SECOND DISPLAY
label SECOND
get xx = x | get yy= y
r = sqrt(xx^2 + yy^2)
phi = atan2(yy, xx) – 0.5 * scale | -- offset
dispt r, phi
```

FIGURE 8.4   Fast Fourier transform of a rectangular pulse. The upper graph shows the real and imaginary parts, and the lower graph shows the absolute value and phase angle. The original displays were in color.

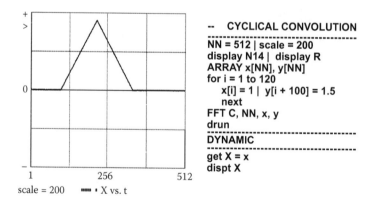

```
--    CYCLICAL CONVOLUTION
-------------------------------------
NN = 512 | scale = 200
display N14 |  display R
ARRAY x[NN], y[NN]
for i = 1 to 120
    x[i] = 1 |  y[i + 100] = 1.5
    next
FFT C, NN, x, y
drun
-------------------------------------
DYNAMIC
-------------------------------------
get X = x
dispt X
```

FIGURE 8.5   Cyclical convolution of two rectangular pulses.

The resulting arrays **xnew** and **ynew** represent the *cyclical convolution* of the real input arrays **x** and **y**. Each element **x[i]** of the original array **x** is replaced with

$$\text{xnew[i]} = \sum_{k=1}^{n} \text{x[k] y[i - k]} \text{ where y[i - n] = y[i]} \ (i = 1, 2, \ldots, n)$$

The cyclical convolution of two real arrays is real, so its imaginary part is zero, and all **ynew[i]** must at least approximately equal 0. This serves as a check on the computation. Figure 8.5 shows a simple example.

REFERENCE

1. Korn, G.A. (2007), *Advanced Dynamic-System Simulation: Model-Replication Techniques and Monte Carlo Simulation*, Wiley, Hoboken, NJ.

Appendix: Simulation Accuracy and Integration Techniques

A.1 SIMULATION ACCURACY AND TEST PROGRAMS

A.1.1 Introduction

Double-precision (64-bit) floating-point number representation resolves over 14 decimal digits, which is much better than most engineering measurements. However, *differential-equation-solving errors* can range all the way from 0.001% for careful trajectory computations to 5% and even 10% for long-running simulations. The choice of an integration rule involves a compromise between accuracy and computing time. Care is needed, for low-accuracy differential equation solving can produce plausible-looking garbage. Because of the enormous volume of simulation-type scientific computation, numerical integration is, after over 200 years of development, still an active research area. A vast literature exists; References 1 through 8 are useful surveys.

Integration errors arise (1) from *local truncation*, that is, from the finite-step approximation to true integration and (2) to *roundoff*, mainly because state-variable accumulation involves addition of many small terms. Errors also affect subsequent integration steps and can thus grow into worse "global" errors.

A.1.2 Roundoff Errors

Using 100,000 integration steps—more than most simulations require— might cause 1 million additions with some integration rules. The resulting percentage roundoff errors depend on the state-variable magnitudes, but double precision (64-bit) floating-point integral accumulation normally eliminates roundoff effects in engineering computations (astronomers may require 64-bits). But 32-bit arithmetic is not enough. Early flight simulators used 32-bit fixed-point arithmetic for derivative computations and 64-bit arithmetic for integration, but now this is neither necessary nor faster.

A.1.3 Choice of Integration Rule and Integration Step Size

In most practical simulations, *derivative computations consume of the order of 10 times as much computing time as the integration routine.* For this reason, elaborate integration routines, which admit larger step sizes for equal truncation errors, can pay for themselves. The choice of an integration rule depends on the specific problem, but an exhaustive study of the possible choices is rarely practical in routine simulation work. To save computing time, one will also want to use the largest value of **DT** consistent with acceptable accuracy. But how can you determine the accuracy of a new simulation time history?

It is also a good idea to check the plausibility of simulation results in light of engineering experience with similar systems. If such experience exists, one can examine the simulation model as well as the solution method.

Different solution methods can sometimes be compared with an analytical solution for a few special cases (Figure A.1). It is often useful to test the sensitivity of simulation results to *changes* in the integration routine and in the integration step **DT** by

- Comparing results for different integration routines

- Decreasing **DT** until a further decrease no longer changes the results

To compare, say, integration Rule 1 (2nd-order Runge–Kutta) and Rule 3 (4th-order Runge–Kutta), program

```
irule 1 | display N15 | drunr
display 2
irule 3 | display N14 | drunr
```

can be employed, where **display 2** keeps time histories of successive runs on the same display, and **display N15** and **display N14** specify curve colors under Windows (Chapter 1, Section 1.3.7). Solution curves result-

-- A SIMPLE ERROR STUDY

```
TMAX = 5 | NN = 101 | DT = 0.02
x = 1
display 0 | -- run without display
drun | -- until t = TMAX
```

```
display 1 | -- now turn listing on
TMAX = 0.2 | NN = 11
drun | -- and continue without reset
-- to measure the error
```

DYNAMIC

```
d/dt x = y | d/dt y = -x
X = cos(t) | ERROR = x - X
type x, X, ERROR
```

t, x, X, ERROR

5.02000e+000	3.03103e-001	3.02783e-001	3.20424e-004
5.04000e+000	3.22102e-001	3.21782e-001	3.19696e-004
5.06000e+000	3.40972e-001	3.40653e-001	3.18825e-004
5.08000e+000	3.59705e-001	3.59387e-001	3.17809e-004
5.10000e+000	3.78294e-001	3.77978e-001	3.16647e-004
5.12000e+000	3.96733e-001	3.96417e-001	3.15339e-004
5.14000e+000	4.15012e-001	4.14698e-001	3.13884e-004
5.16000e+000	4.33125e-001	4.32813e-001	3.12282e-004
5.18000e+000	4.51066e-001	4.50755e-001	3.10534e-004
5.20000e+000	4.68825e-001	4.68517e-001	3.08637e-004
5.22000e+000	4.86397e-001	4.86091e-001	3.06593e-004
5.24000e+000	5.03775e-001	5.03471e-001	3.04401e-004

FIGURE A.1   A program for checking the solution of the harmonic oscillator problem

d/dt x = xdot | d/dt xdot = - x with **x(0) = 1**

against its analytical solution **x = cos(t)**. Since errors become more pronounced for large values of **t**, the **display 0** statement causes the program to run without display for **Tacc** time units, and then produces a listing using **type**. The first listing is for 2nd-order Runge–Kutta integration, and the second listing is for 4th-order Runge–Kutta integration. *(Continued)*

5.02000e+000	3.02783e-001	3.02783e-001	-6.41195e-009
5.04000e+000	3.21782e-001	3.21782e-001	-6.39766e-009
5.06000e+000	3.40653e-001	3.40653e-001	−6.38048e-009
5.08000e+000	3.59387e-001	3.59387e-001	-6.36040e-009
5.10000e+000	3.77978e-001	3.77978e-001	-6.33742e-009
5.12000e+000	3.96417e-001	3.96417e-001	-6.31151e-009
5.14000e+000	4.14698e-001	4.14698e-001	-6.28267e-009
5.16000e+000	4.32813e-001	4.32813e-001	-6.25089e-009
5.18000e+000	4.50755e-001	4.50755e-001	-6.21616e-009
5.20000e+000	4.68517e-001	4.68517e-001	-6.17849e-009
5.22000e+000	4.86091e-001	4.86091e-001	-6.13786e-009
5.24000e+000	5.03471e-001	5.03471e-001	-6.09427e-009

FIGURE A.1 (Continued)   A program for checking the solution of the harmonic oscillator problem.

$$d/dt\ x = xdot\ |\ d/dt\ xdot = -x\ \text{with}\ x(0) = 1$$

against its analytical solution $x = \cos(t)$. Since errors become more pronounced for large values of t, the display 0 statement causes the program to run without display for Tacc time units, and then produces a listing using type. The first listing is for 2nd-order Runge–Kutta integration, and the second listing is for 4th-order Runge–Kutta integration.

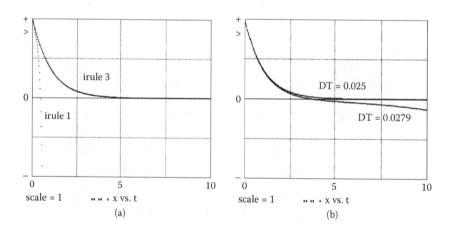

FIGURE A.2   Successive simulation runs let you judge the effects of (a) different integration rules and (b) different integration step sizes. The original displays showed different curves in different colors.

ing from **irule 1** will be white, and curves resulting from **irule 3** will be yellow (Figure A.2a).

A similar routine tests the effect of decreasing the integration step DT:

display N 15 | drunr

display 2
DT = 0.1 * DT | display N 14 | drunr

(Figure A.2b). Such simple tests can be added to many Desire programs, as in the user example **bouncer.src** on the book CD. Remember, though, that you are testing the sensitivity of the solution to integration rule or **DT** changes, not the absolute solution accuracy.

A.2 INTEGRATION RULES AND STIFF DIFFERENTIAL EQUATION SYSTEMS

A.2.1 Runge–Kutta and Euler Rules

Table A.1 lists Desire's integration rules. When a program containing differential equations does not specify the integration rule, it defaults to Rule 1 (2nd-order Runge–Kutta–Heun Rule), a simple and fast routine which does quite well in many practical simulations. Rule 3 (4th-order Runge–Kutta) normally has substantially smaller truncation errors for the same value of **DT**.

Variable-step Runge–Kutta rules (Rules 4, 7, and 8) compare two Runge–Kutta formulas of different order. The step size doubles when the **absolute** difference is less than **ERMIN**, until **DT** reaches **DTMAX**. If the system variable **CHECKN** is a positive integer **n**, the step size **DT** is halved when the absolute difference of the two expressions for the **n**th state variable exceeds **ERMAX**. For **CHECKN = 0**, **DT** is halved when the *relative* difference exceeds **ERMAX** for *any* state variable. A variable-step deadlock error results when **DT** attempts to go below **DTMIN**; the deadlocked difference can then be read **in ERMAX**. Note that **ERMAX** is a bound on the Runge–Kutta test difference, *not* on the solution accuracy.

If not specified, DTMAX defaults to COMINT = TMAX/(NN - 1), DTMIN defaults to DT/16, ERMAX defaults to 0.001, and ERMIN defaults to ERMAX/100.

Rule 4 (Runge–Kutta 4/2) is the only variable-step Runge–Kutta rule we normally use. The lower-order Rules 7 and 8 may not justify the variable-step overhead. Variable-step Runge–Kutta routines save time by rejecting unnecessarily small **DT** values, but they may accept **DT** values too large

TABLE A.1   Desire Integration Rules

(a) EULER AND RUNGE–KUTTA RULES

(up to 40,000 state variables)

k1 = G(x, t) * DT

RULE 1 (*fixed-step 2nd-order Runge–Kutta–Heun rule; this is the default rule*)

k2 = G(x + k1, t + DT) * DT

x = x + (k1 + k2)/2

RULE 2 (*fixed-step explicit Euler rule, 1st order*)

Users may change $_{DT}$ in the course of a simulation run.

x = x + G(x, t) * DT = x + k1

RULE 3 (*fixed-step 4th-order Runge–Kutta rule*)

Users may change **DT** in the course of a simulation run.

k2 = G(x + k1/2, t + DT/2) * DT k4 = G(x + k3, t + DT) * DT

k3 = G(x + k2/2, t + DT/2) * DT

x = x + (k1 + 2 * k2 + 2 * k3 + k4)/6

Variable-Step Runge–Kutta rules compare two Runge–Kutta formulas of different order. The step size doubles when the *absolute* difference is less than **ERMIN**, until **DT** reaches **DTMAX**. If the system variable **CHECKN** is a positive integer *n*, then the step size **DT** is halved if the absolute difference of the two expressions for the *n*th state variable exceeds **ERMAX**. If **CHECKN** = 0, then **DT** is halved when the *relative* difference exceeds **ERMAX** for *any* state variable. A variable-step deadlock error results if **DT** attempts to go below **DTMIN**; the deadlocked absolute difference can then be read in **ERRMAX**.

RULE 4 (*variable-step Runge–Kutta 4/2*) compares the 4th-order Runge–Kutta result with

x = x + k2

RULE 5 (*2nd-order R–K–Heun,* like **RULE 1** but users may change **DT** during a run)

RULE 6 (spare, not currently implemented)

RULE 7 (*variable-step Runge–Kutta 2/1*) compares

k2 = G(x + k1, t + DT)

x = x + (k1 + k2)/2 with x = x + k1

RULE 8 (*variable-step Runge–Kutta–Niesse*) compares

k2 = G(x + k1/2, t + DT/2) * DT

k3 = G(x - k1 + 2 * k2, t + DT) * DT

x = x + (k1 + 4 * k2 + k3)/6 with x = x + (k1 + k3)/2

(b) ADAMS-TYPE VARIABLE-ORDER/VARABLE-STEP RULES

(up to 1000 state variables)

RULE 9 functional iteration

RULE 10 chord/user-furnished Jacobian

RULE 11 chord/differenced Jacobian

RULE 12 chord/diagonal Jacobian approximation

(*Continued*)

TABLE A.1 (Continued)   Desire Integration Rules

(c) GEAR-TYPE VARIABLE-ORDER/VARABLE-STEP RULES

(for stiff systems, up to 1000 state variables)

RULE 13 functional iteration

RULE 14 chord/user-furnished Jacobian

RULE 15 chord/differenced Jacobian

RULE 16 chord/diagonal Jacobian approximation

For Integration Rules 9 through 16 the experiment protocol must specify a maximum *relative* error **ERMAX** for all state variables. Values set to 0 are automatically replaced by 1 (user examples **orbitx.src t, to22x.src, rule15.src**). The initial value of **DT** must be low enough to prevent integration-step lockup.

For Integration Rules 10 and 14 the experiment-protocol script must declare an **n**-by-**n** *Jacobian matrix* **J[l, k]** whose elements are defined by assignments in a compiled **JACOBIAN** *program segment* (user examples **orbitx.src, fuel.src, fuel3.src**).

---

for acceptable accuracy. For this reason, Desire lets you set a maximum value **DTMAX** of **DT**. You can try to decrease **DTMAX** until a further decrease does not affect the solution.

A.2.2 Implicit Adams and Gear Rules

Table A.1 lists the implicit variable-order/variable-step Adams and Gear rules furnished with Desire. These are C routines based on the public-domain DIFSUB, LSODE, and LSODI Fortran packages [2,5,7]. Rules 9 through 12 are implicit Adams-type predictor/correctors, and Rules 13 through 16 are Gear-type rules. All these integration rules try to adjust the order as well as the step size of the integration routine to reduce truncation errors. The Gear routines are specifically designed for use with stiff differential equation systems (Sections A.2.3 and A.2.4). References 2, 5, and 7 describe the integration routines and discuss relevant theory.

A.2.3 Stiff Differential Equation Systems

A differential equation system

$$d/dt\ x_i = f_i(x_1, x_2, \ldots ; t) \quad (i = 1, 2, \ldots)$$

is called **stiff** when the eigenvalues of its Jacobian matrix $[\partial f_i/\partial x_k]$ differ widely in absolute value. Roughly speaking, these eigenvalues relate to reciprocals of system response time constants. As an example, the simple linear system

$$\text{d/dt } x = y \quad \text{d/dt } y = -100 * x - 101 * y \quad \text{(A.1)}$$

has solutions

$$x = A\, e^{-t} + B\, e^{-100t} \quad y = -A\, e^{-t} - 100\, B\, e^{-100t} \quad \text{(A.2)}$$

corresponding to its Jacobian eigenvalues (system time constants) −1 and −100. The second term in each solution (Equation A.2) decays so quickly that the solutions look similar to those of the first-order differential equation **d/dt x = - x**, which could be solved fairly accurately with an integration step **DT** as large as 0.1. However, a Runge–Kutta solution of our stiff second-order system is not only inaccurate but *invalid* unless we reduce **DT** below about 0.01 to match the smaller of the two eigenvalues; this is apparent in Figure A.2b. Quite generally, "explicit" Runge–Kutta rules require **DT** to be of the order of magnitude of the smallest Jacobian eigenvalue, even though the actual system response seems to be much slower.

In such situations, implicit-type integration rules designed for stiff systems [1,2,5,7] can save computing time or restore stability. Unfortunately, such rules require numerical inversion of the Jacobian matrix, which can be large and must be either user-supplied or computed by approximate differentiation. Integration rules 13 through 16 (Table A.1) are implicit-type variable-order/variable-step rules suitable for stiff-system integration.

For small problems and short simulation-run times, it may be preferable to simply use small **DT** values, since roundoff-error accumulation is rarely a problem with 64-bit computations. Sometimes an attractive way to deal with stiff systems is to modify the model so that the state equations responsible for the fastest system time constants are replaced by defined-variable assignments relating system variables. One might, for instance, approximate the solution (Equation A.2) of the stiff system (Equation A.1) by replacing the second differential equation with the assignment **y = - x**.

A.2.4 Examples Using Gear-Type Integration

The satellite orbit computation in Chapter 2, Section 2.3.5, works even with simple Euler integration when **DT** is small enough. But variable-step integration rules produce accurate results much more quickly, since one can safely reduce the integration step **DT** substantially for parts of the elliptical orbit in Figure 2.15 (Chapter 2). The program in Figure 2.15 specifies Gear integration with the Jacobian computed through built-in numerical differentiation (**irule 14**). To solve the same problem with a *user-supplied Jacobian matrix*, substitute **irule 15** for **irule 14** in the experiment protocol and declare the Jacobian **J** with

ARRAY J[4, 4]

Compute the required partial derivatives **J[i, k]** by differentiating the given derivative assignments

d/dt r = rdot   d/dt rdot = - GK/(r * r) + r * thdot * thdot
d/dt theta = thdot d/dt thdot = 2 * rdot * thdot/r

and add the lines

JACOBIAN
J[1,3] = 1
J[2,4] = 1
J[3,1] = 2 * GK/(r * r * r) + thdot * thdot
J[3,4] = 2 * r * thdot
J[4,1] = 2 * rdot * thdot/(r * r)
J[4,3] = 2 * thdot/r
J[4,4] = 2 * rdot/r

at the end of the DYNAMIC program segment. All other Jacobian elements (partial derivatives) **J[i, k]** simply default to 0. Figure A.3 shows how the variable integration step **DT** increases and decreases periodically with the radius vector **r** of the satellite trajectory. Note that the satellite velocity is greatest when **r** is small (i.e., near the earth). You can try the same experiment with other variable-step rules.

The user examples **fuel2.src** and **fuel3.src** on the book CD illustrate the solution of another simulation program with stiff differential equations (xenon poisoning in a nuclear reactor) and a much more complicated Jacobian. Both **irule 14** and **irule 15** produce good results.

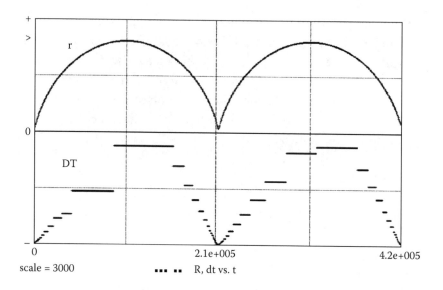

FIGURE A.3   Rescaled time histories of the radius vector and the integration step for the satellite orbit problem in Figure 2.14 (Chapter 2), using Gear integration (Rule 15).

A.2.5 Perturbation Methods Can Improve Accuracy

Consider a state equation

$$d/dt\ x = G(x, t, a) \tag{A.3}$$

where $a = a(t)$ is a variable affecting the solution. The following theory applies equally well when x, G, and a are vectors, and Equation A.3 represents a system of differential equations.

To ascertain the effects of small changes (perturbations) $\delta a(t)$ in $a(t)$ on the perturbed solution $x(t)$, we write

$$a = a0(t) + \delta a(t) \quad x = x0(t) + \delta x(t) \tag{A.4}$$

where $a0(t)$ is the solution of the simpler system

$$d/dt\ x0 = G(x, t, a0) \tag{A.5}$$

which is assumed to be accurately known either as an analytic solution, or from an earlier accurate computation.

We can now find the corrections **δx(t)** caused by the perturbations **δa(t)** by solving

$$\text{d/dt } \delta x = G[x0(t) + \delta x, t, a0(t) + \delta a(t)] - G[x0(t), t, a0(t)] \quad (A.6)$$

When **|δx|** is small compared to **|x|**, errors in **δx** have a relatively small effect on the solution **x(t) = x0(t) + δx(t)**. With a little luck, *the accuracy improvement can be dramatic*. In particular, it may be possible to replace the exact perturbation equation (Equation A.6) with a Taylor series approximation

$$\text{d/dt } \delta x = (\partial G/\partial x)\, \delta x + (\partial G/\partial a)\, \delta a$$

where the partial derivatives are known functions of the unperturbed solution **x0(t)** and the time **t**, but are independent of **δx** and **δa** (*linearized perturbation equation*). This is, in fact, how we derived the linearized aircraft flight equations in Chapter 2, Section 2.3.2. Perturbation techniques are especially useful for computing aerospace vehicle trajectories [9].

A.2.6 Avoiding Division

Since division usually takes more time than multiplication, it may help to replace a differential equation of the form

$$\text{dy/dx} = F(x, y)/G(x, y)$$

with a pair of differential equations

$$\text{d/dt } x = a\, G(x, y) \qquad \text{d/dt } y = a\, F(x, y)$$

with **x(0) = 0**.

REFERENCES

1. Cellier, F., and E. Kofman (2006), *Continuous-System Simulation*, Springer, New York.
2. Gear, C.W. (1971), DIFSUB, Algorithm 407, *Comm. ACM*, 14, No. 3, 3/7.
3. Gupta, G.K. et al. (1985), A Review of Recent Developments in Ssolving ODEs, *Computing Surveys*, 17:5–47, March.
4. Hairer, E. et al. (1987), *Solving Ordinary Differential Equations* (2 Vols.), Springer, Berlin.

5. Hindmarsh, A.C. (1980), LSODE and LSODI, *ACM/SIGNUM Newsletter*, 15, No. 4.
6. Lambert, J.D. (1991), *Numerical Methods for Ordinary Differential Equations: The Initial-Value Problem*, John Wiley & Sons, New York.
7. Shampine, L.F., and H.A. Watts (1984), Software for ordinary differential equations, in L.R. Crowell (Ed.) *Mathematical Software*, Prentice-Hall, Englewood Cliffs, NJ.
8. Butcher, J.C. (1980), *The Numerical Analysis of Ordinary Differential Equations*, Wiley, Chichester, U.K..
9. Cellier, F. (2010), *Numerical Simulation of Dynamic Systems*, Springer, New York.
10. Fogarty, L.E., and R.M. Howe (1982), Analog-Computer Solution of the Orbital Flight Equations, *IRETEC*, August.

Index

1st order differential equations, 3–4. *See
also* State equations

A

Adams rules, 189
Aerospace and related application
    examples
    pilot ejection study, 159–163
    satellite roll-control simulation, 92–95
Aerospace and related applications
    examples
    autopilot simulation, simplified, 37–40
    ballistic trajectories, 32–34
    simple flight simulation, 35–37
    torpedo trajectory simulation, 38–41
Aircraft, simulation of motion of, 35–36
Amplitude modulation, 88
Amplitude/phase frequency response, 126
Analog bandpass filter simulation,
    127–128
Analog data, simulating reconstruction
    of, 64
Analog plant, simulation of with digital
    PID controller, 64–67
Analog systems, programming linear-
    system transfer functions for,
    125–126
Array-value plotting, 173–174
Arrays
    equivalent, 104–105
    filling with data, 105–107
    manipulation of using DYNAMIC-
        segment code, 174–175
    simple declarations, 103–104
    simultaneous transformation of,
        180–181
    STATE, 105

time history function storage in,
    138–144
Automatic display scaling, 140–141
Autopilot simulation
    simple program for, 39
    simplified, 37–40

B

Ballistic trajectory simulation, 32–34
Bang-bang control system, simulation of
    using submodels, 92–97
Bouncing ball simulation, 146–149
Box–Muller-type formula, 78
Branching, 99–100
    conditional, 100–101
Breakpoint abscissas, 70

C

Cannonball trajectories. *See also* Ballistic
    trajectory simulation
    simulation program for, 33
Case sensitivity, 21
Channels, 108
Chemical reactions, use of population-
    dynamics models to study, 47
City-block norm, 131
Classical dynamic system models, 23–24
clear statements, 104
Closing files or devices, 109
Code, difference equation *vs.* differential
    equation, 61–63
Combined systems, sample/hold
    operations and, 61–64
Command Window, 3
Communication interval, 7
Communication points, 7

Comparators, 77
  hysteresis models and, 84, 86
COMPLEX, 176–177
Complex frequency-response plots, 179
Complex numbers
  operations, 177–178
  plots, 178–179
Conditional branching, 100–101
Conformal mapping, 179
connect statements, 108
Console input, interactive, 109
Console output, 107–108
Continuous functions, simple limiters
    and, 74
Control system design, 49
  elaborate models of, 56
Control system noise, elaborate plant
    models and, 56
Control systems
  sampled-data, 64–67
  simulation of using submodels, 92–97
  transfer functions and frequency
    response, 56–57
Coupled oscillators
  example model of using submodels,
    90–92
  matrix/vector model of, 121–122
Cyclical convolutions, 181–183

D

Damping coefficients, 135
data lists, 106–107
deadc(x), 77
Deadspace comparator, hysteresis models
    and, 84–85
Deadspace function [deadz(x)], 74
deadz(x), 74
Defined variables, 60
  differential equation systems with, 4–5
Defined-variable assignments, 60–61
delay, 141–143
Delay line circuit, simulation of, 29–32,
    124–125
Derivative routines, 6, 61–63
Desire, 1–3
  getting started with, 13–14
  installation of, 12

integration rules, 190–191
Device input, 109–110
Device output, 108–109
Devices, closing, 109
Difference equation code, combination of
    with differential equation code,
    61–63
Difference equations, 3
  control system models using, 57–58
  models using, 57–58
  primitive, 59–60
  programming using, 59–64
  vector, 131–132
Difference-equation state variables, 60–61.
    See also State variables
Differentiable approximations, 79
Differential equation code, combination
    of with difference equation
    code, 61–63
Differential equation systems with defined
    variables, 4–5
Differential equations, 3
  higher-order, 23–24
  Lorenz's nonlinear, 28
  solving errors, 185
  stiff systems of, 189–195
  Van der Pol's, 24–26
Digital controller, simulation of an analog
    plant with, 64–67
Digital filters, programming linear-system
    transfer functions for, 126–129
disconnect statements, 109
Discontinuities, integration through,
    79–80
Display scaling, 17–18
  automatic, 140–141
display statement, 15
DOT products, 129–130
DOT statements, 117
drun statement, 6, 111
drunr statement, 15, 22, 61, 111
DT, 15, 22, 186–188
  decreasing, 95–96
DTMAX, 80
Dummy arguments
  in FUNCTION declarations, 73
  in submodels, 90
Dummy integration, 172

dump command, 112
DYNAMIC program segments, 5
    fast array manipulation using, 174–175
    fast graph plotting using, 172–174
    multiple, 110–111, 159–163
    statistical computations using, 175–176
    vector expressions in, 118–119
DYNAMIC statement, 15
DYNAMIC system models, vectors and
        matrices in, 118–122
Dynamic system models, 57–58
Dynamic-system models, simulation
        programs and, 1–5

E

Ecology, host–parasite problem
        simulation, 45–46
edit, 22
Editor Windows, 3
    multiple, 14
Electrical circuit simulation, 29, 31–32,
        124–125
Electronic-switch benchmark problem,
        152–153
else clauses, 100–101
end while statements, 101
Epidemic propagation, simulation of,
        43–45
Equivalent arrays, 104–105
    declaration of, 132–133
Error correction, interactive, 111
Error measures, test inputs and, 51–52
Error-rate feedback, 56
Euclidean norms, 130
Euler integration, 9
EUROSIM benchmark problems, 145–146
    electronic switch, 152–153
    lithium cluster, 147–148
    peg-and-pendulum, 149–151
Exercise models, 5–11
exit statements, 101
Experiment protocol, 5
    debugging, 111–112
    output and input, 107–111
    procedures, 101–103
    programs and commands, 21–23

vectors and matrices in scripts,
        116–118
Experimental-protocol scripts, 5
Explicit Euler rule, 9
Explicit integration rules, 11
Expressions, evaluation of, 21
Extra state variables, 52

F

f(), 119
Fast Fourier transforms, 180–182
Fast graph plotting, 172–174
Feedback position-control system. See
        Servomechanisms
File input, 109–110
File output, 108–109
Files, closing, 109
Flight simulation
    linearized flight equations, 36–37
    pitch-plane flight equations, 35–36
for loops, 101
Forrester-type system dynamics, 160,
        164–169
Fourier transformations, 180–182
Frequency modulation of waveforms,
        87–88
Frequency-response function, 57
    plots, 178–179
Full-wave rectifier, 75
Function generation
    general-purpose, 69–74
    using function tables, 70–73
Function plot, 172–173
Function switch, 77
Function tables, 70–73
Functions
    storage and recovery of with store and
            get, 138–141
    user-defined, 73–74

G

gain, 88
gauss(0), 78–79
Gaussian noise, 78
Gear backlash, 83
Gear rules, 189

Gear-type integration, 192–193
get statements, 111, 138–141
Glucose tolerance test, simulation of, 153–156
Gradient-measuring simulation run, 54
Graph plotting, 172–174
Graph Window, 3
Gyro simulation, 37–38

H

Hamming norms, 131
Hard impact simulation, 146–149
Help facilities, 113
Hexidecimal integer conversions, 177
Host-parasite problem simulation, 45–46
Human blood circulation, 154, 156–159
Hysteresis, models with, 83–85

I

Identity matrices, 116
if statements, 23
    conditional branching and, 100–101
Implicit Adams rules, 189
Implicit Euler rule, 11
Implicit Gear rules, 189
Implicit rules, 11
Impulse response, 126
Index-shifted vectors, 122–124
    replacement of derivative assignments by, 124–125
Inductance/capacitance delay line, simulation of, 124–125
Inner products, 129–130
input statements, 109–110, 123–124
INTEGER, 176–177
Integer conversions, 177
Integral control, 56
Integration, 6–7
    discontinuities and, 79
    Gear-type, 192–193
    numerical, 9–11
    step operator and, 80
    through discontinuities, 11
Integration errors, 185
Integration rules, 8, 22, 186–188

stiff differential equations systems and, 188–195
Integrator feedback, 86
Integrators, output-limited, 76
Interactive commands, 22
    console input, 109
Interactive error correct, 111
Interactive modeling, 1–2
    multiple Editor Windows, 14
    wish list for, 11–12
Interpolation, 64, 70–71
Interpreter graphics, 178
Inverse functions, generation of, 88–89
Invocation variables, 90
Iterative parameter optimization, 53–56

J

Jacobian eigenvalues, 79, 191

K

keep command, 14, 111

L

label statements, 110–111
Labels, declaration of, 99–100
Library functions, 69–70
    complex number operations and, 177–178
    Piecewise-linear, 74–75
lim(x), 74
lim[tri(x)], 74
Limit cycle oscillation modeling, 25
Limiter functions, useful relations between, 75–76
Limiters
    output, 76
    simple, 74–75
Linear control systems, experiment-protocol scripts for, 57
Linear controller definition, 50
Linear equations, matrix inversion and solution of, 118
Linear systems
    matrix/vector models of, 121–122

programming transfer functions, 125–126
Linearized flight equations, 36–37
Linux, installing Desire on, 12–13
Lithium-cluster-benchmark simulation program, 147–148
Local truncation, 185
Logarithmic plotting, 145–146
Loops, 101
Lorenz's differential equations, 28

M

Mackey-Glass time series, 143–144
Matrices, 103–104, 132–133
    inversion of and solution of linear equations, 118, 121
    null and identity, 116
    rotation, 120–121
    submodels with, 133–134
    sums and products, 117
    transposition of, 117
MATRIX, 117–118
Matrix operations, dynamic-segment, 132–133
Matrix/vector models of linear systems, 121–122
Matrix/vector products, 119–120
Maximum absolute error, 52
Maximum functions, 76
Maximum squared error, 52
Maximum/minimum tracking, 81–83
Minimum functions, 76
Model parameters, 4, 15
Model replication, 134–138
Models, 1
Modified triangle function (lim[tri(x)]), 74
Monte Carlo simulation, vectorized, 137
Multiple Editor Windows, interactive modeling with, 14
Multiple runs
    crossplotting results from, 159–163
    splicing of complicated time histories using, 146–153
Multirun simulation studies, 5–6
Multistep rules, 10

N

Nested if statements, 100–101. *See also* if statements
Neural network simulation, 137–138
NN, 15, 22
Noise generators, 77–79
Nonlinear controllers, 56
Nonlinear oscillators, phase-plane plots and, 24–28
Norms
    Euclidean, 130
    hamming, 131
    taxicab, 131
Notebook file, 112–113
Nuclear reactor simulation, 28–30
Null matrices, 116
Numbered channels, 108
Numerical integration, 9–11, 185
Nyquist plots, 179

O

Octal integer conversions, 177
One-dimensional arrays, 103. *See also* Vectors
Operating points, 29
Optimization gain, 88
OUT statement, 8–9, 16, 61–62, 80
Output requests, 7
Output timing, 7–9
Output-limited integrators, 76
Output-rate feedback, 56

P

Parameter optimization
    iterative, 53–56
    parameter-influence studies, 52–53
    test inputs and error measures, 51–52
Parameter-influence studies, 52–53
    model replication and, 135–136
Peg-and-pendulum benchmark problem, 149–151
Pendulums
    simulation of, 26–28
    simulation of EUROSIM peg-and-pendulum problem, 149–151

Periodic sampled-date operations, 62
Periodic signals, program for digital
    generation of, 86
Perturbation methods, 193–195
Phase-modulation, 88
Phase-plane plots, nonlinear oscillators
    and, 24–28
Phugoid oscillation, 37
PHYSBE model of human blood
    circulation, 154, 156–159
Physics examples
    dynamic system models and higher
        order differential equations,
        23–24
    electrical circuit simulation with 1401
        differential equations, 29, 31–32
    nonlinear oscillators and phase-plane
        plots, 24–28
    simple nuclear reaction simulation,
        28–30
Physiological models, 153–159
Piecewise-linear library functions, 74–75
Pilot ejection problem, crossplotting for,
    159–163
Pitch-plane flight equations, 35–36
Plant models, control system noise and, 56
Plotting, 172–174
Point-by-point plotting, 178
Population dynamics
    generalizations for models of, 45–47
    host-parasite problem simulation,
        45–46
    simulation of epidemic propagation,
        43–45
Primitive difference equations, 59–60
Probabilities, computation of using
    Desire, 175
Procedures, experiment-protocol, 101–103
proceed statements, 100–101
Program control, 99–103
Proportional error feedback, 56
Proportional/integral/derivative (PID)
    controller simulation, 64–67
Pseudorandom noise, 78–79
    sampling, 59
Publication copy preparation, 19

Q

q, 59, 61, 80–81
Quantization, 77–78

R

ran(), 77, 79
read assignments, 106–107
Real-time flight simulation, 38
Recurrence relations, 59, 80–89
Relay comparator, 77
repeat loops, 101
Rescaling, 17–18
reset, 61
restore statements, 107
Roll displacement, 93
Root-locus plots, 179
Rotation matrices, 120–121
round(x) function, 70
Roundoff errors, 185–186
Runge–Kutta rules, 10, 186, 188–189

S

Sample averages, computation of using
    Desire, 175
SAMPLE m statement, 61–62, 80
Sample/hold operations, control systems
    and, 61–64
Sampled data, 57–59
    operations with, 8–9
    periodic operations, 62
    simulation of reconstruction of, 64
    transferring, 64
Sampled periodic noise, 79
Sampled-data control system simulation,
    64–67
Sampled-data operations, 58–59
Sampling points, 7
Sampling rate, 59
sat(x), 74
Satellite orbit
    simulation, 41
    simulation program, 43
Saturation limiters, 74
Saving programs, 13–14
Sawtooth waveforms, 85–87

Schmitt trigger, 84
  integrator feedback around, 85–86
Script debugging, 112
Servomechanisms
  bang-bang controller model, 92–97
  modeling of with iterative parameter
    optimization, 53–56
  simulation of, 49–51
sigmoid(x) function, 70
sign(x), 76
Signal generators, 85–89
Signal modulation, 87–88
Signal quantization, 77–78
Simple backlash, hysteresis and, 83–84
Simple limiters, 74–75
Simple pendulum, simulation of, 26–28
Simulation, 1
  complete program for, 14–15
Simulation parameters, 7, 15
Simulation programs
  Aerospace and related application
    examples, 32–41, 92–95
  analog bypass filter, 127–128
  analog plant with digital PID
    controller, 64–67
  bouncing ball, 146–149
  control systems, 92–97
  delay-time circuit, 29–32, 124–125
  dynamic-system models and, 1–5
  electronic-switch benchmark problem,
    152–153
  entering and running, 13–14
  exercise models, 5–11
  glucose tolerance test, 153–156
  lithium-cluster-benchmark problem,
    147–148
  Mackey-Glass time series, 143–144
  peg-and-pendulum benchmark
    problem, 149–151
  PHYSBE model of human blood
    circulation, 154, 156–159
  Physics examples, 23–32
  pilot ejection study, 159–163
  population dynamics examples, 43–47
  servomechanisms, 49–51
  time-delay, 141–143
Simulation runs, 5
  numerical integration, 9–11

output timing, 7–9
  pausing, 14
  updating of variables, 6–7
Simulation studies
  multirun, 5–6
  parameter optimization and, 51–56
Sine-function table, 72
Single-stepping, 112
Square matrices, 116
  inversion of, 118
Square wave waveforms, 85–87
stash statements, 18
STATE arrays, 105
STATE declaration, 90
State equations, 3–4
State variables, 3, 15, 52, 60
  invoked, 90
  subscripted, 105
State vectors, 105
Static models, 3
Statistical computations using Desire,
  175–176
Step operators, integration and, 80
step statement, 61–62, 79–80
Stiff differential equation systems,
  189–195
STOP statements, 22, 112
store statements, 111, 138–141
Stripchart-type displays, 17–18
SUBMODEL declaration, 89–90
Submodels, 89
  bang-bang control system simulation
    using, 92–97
  coupled-oscillator problem using,
    90–92
  declaration of, 89–90
  invocation of, 90
  with vectors and matrices, 133–134
Subscripted state variables, 105
Subscripted variables, 104
  filling arrays using, 105
Switch function, 76
switch(x), 76
Switches, 76–77
System design, clarification of using
  submodels, 89–92
System dynamics, 160, 164–169

T

Taxicab norms, 131, 133–134
tdelay, 141–143
term statements, 7, 23
Test inputs, error measures and, 51–52
thedot, 27
then clauses, 100–101
theta, 26–27
tim(0) function, 70
Time averages, computation of using
    Desire, 175
Time delay simulation, 141–143
time statement, 22
Time-history output
    display scaling and stripchart-type
        displays, 17–18
    programming graphs and listings, 17
    storage and printing, 18–19
Time-history sampling, 7–8
TMAX, 15, 22, 58
Torpedo trajectory
    simulation, 38–41
    simulation program, 42
trace statements, 112
Track/hold circuits, 81–83
Transfer-function input/output relations,
    56–57
Translunar satellite orbit
    simulation, 41
    simulation program, 43
tri(x), 74
Triangle function [tri(x)], 74
Triangle waveforms, 85–87
TRIGA nuclear reactor, simulation of,
    28–30
trnc(x) function, 70
Two-dimensional arrays, 103. *See also*
    Matrices
type statements, 18

U

u(t), 49, 51–52

V

Unit pulse, 77
until statements, 101
Updating assignments, 60
User-defined functions, 73–74

V

Van der Pol's differential equation,
    simulation of, 24–26
Variable-step integration rules, 11
Variables
    functions of one, 70–71
    functions of two, 71–72
    updating of, 6
Vector assignments, replication of models
    with, 134–135
Vector difference equations, 131–132
Vector expressions, 118–119
    Matrix/vector products in, 119–120
Vector norms, 130–131
Vector products, 132
Vectorized Monte Carlo simulation, 137
Vectorized parameter-influence study, 136
Vectors, 103
    index-shifted, 122–124
    state, 105
    submodels with, 133–134
    sums and products, 117

W

Waveforms, 85–87
while loops, 101
Width-modulated pulses, 88
Windows, installing Desire on, 12–13
Working run, 54
World simulation, 164–169
write statements, 107–108
write x statement, 22

Y

Yaw, simulation of, 38–39